ENDO

"I have long considered synchronicities as profound signs from the universe, affirming that we are on the right path. "The Playful Universe—Synchronicity and the Nature of Consciousness" masterfully normalizes what has been, until now, often regarded as a fringe concept, showcasing its vital importance in a post-materialist world. Through a rich tapestry of insights and personal anecdotes, this book wonderfully highlights how embracing synchronicities can significantly support one's journey of intuitive development. I highly recommend this enlightening read for anyone looking to deepen their understanding of the interconnected nature of consciousness."

Helané Wahbeh, PhD
Director of Research, Institute of Noetic Sciences

"This book provides the deepest exploration into the nature and ramifications of synchronicity, since Carl Jung and Wolfgang Pauli published their original monographs on the topic in 1952."

Jeffrey Mishlove, PhD,
Host and Producer, New Thinking Allowed YouTube channel

"This inspiring and important book shows that there are hidden patterns behind the events of our lives. Containing many fascinating cases of synchronicity—along with the theoretical background to make sense of the cases—the book makes it clear that we live in a meaningful interconnected world."

Steve Taylor PhD,
Author of The Adventure, Time Expansion Experiences and many other books

"A brilliant, multifaceted kaleidoscope of contributions that highlight an essential ingredient of postmaterialist thinking: Efficient causation is not the only means that governs processes in the universe. There is more. And this "more" is spontaneous, unpredictable, mind-related - and yes, playful."

Michael Nahm, PhD,
Institute for Frontier Areas of Psychology and Mental Health, Freiburg, Germany

The Playful Universe is a great romp through infinite fields of possibility. Both intriguing and thought provoking, this collection of musings on synchronicity by multiple authors is a delightful addition to the age-old art of discernment. Is that happenstance meaningful? A breadcrumb from beyond? Coincidence? Or alas… just my ego asserting its preferences.

Joan Borysenko, PhD
Co-author of Your Soul's Compass: What is Spiritual Guidance?

Of all the phenomena which suggest that the universe is driven not by blind, mechanistic forces but by meaning and consciousness, I would place synchronicity as the most potent. The reason for my view is expertly conveyed in the rich collection of writings in this book. The intermingling of personal anecdotes of synchronicities with deep reflections on their significance makes this a highly readable work and one which contributes profoundly to our understanding of the universe in which we live. As Butzer puts it in her chapter, "[W]e live in an intelligent universe, a universe that has infinite creative potential, and that wants to co-create with us, possibly as a way to better know itself."

B Les Lancaster, PhD
Co-Founding Director, Alef Trust; Emeritus Professor of Transpersonal Psychology,
Liverpool John Moores University

In Indigenous Māori thinking, synchronicities are ever-present. Inherently spiritual in nature, such synchronicities are known by the term 'tohu', which translates more readily as sign or symbol. To interpret an event or phenomenon as a tohu is to endorse the idea that seemingly unrelated events are inextricably linked across time and space, bringing meaning and purpose to the interpreter. Although common to a Māori worldview, the concept of synchronicity still struggles for endorsement in contemporary science. Yet, documenting the synchronistic experiences of some of the most well-regarded thinkers of our time, highlights that synchronicity remains meaningfully potent in the lives, and indeed the careers, of many who contribute to the advancement of science; making this an enlightening volume for anyone wanting to know more about what synchronicity.

Natasha Tassell-Matamua, PhD
Associate Professor, Massey University, New Zealand

Profoundly insightful and intellectually stimulating, The Playful Universe masterfully explores how synchronicities—those 'meaningful coincidences'—can transform lives. The authors share their personal journeys, detailing how these unexpected alignments inspired them to embrace a playful collaboration with the universe. These experiences sometimes prompted significant changes, such as career shifts or relocations, that ultimately led them to more fulfilling paths. By blending intuition with logic, the authors also arrive at a worldview that integrates science and spirituality, while demonstrating how paying attention to the nuances of life can unveil its inherent magic.

Diane Hennacy Powell, MD
Author of The ESP Enigma: A Scientific Case for Psychic Phenomena

The Playful Universe

Synchronicity and the Nature of Consciousness

Volume V
Postmaterialist Sciences Series

Edited by
Marjorie Woollacott, PhD
David Lorimer
Gary E. Schwartz, PhD

AAPS
PRESS

AAPS Headquarters, Tucson, AZ.
AAPS Press and Information on AAPS,
P.O. Box 156,
Battle Ground, WA 98604

L.C. Cat. No.: 1-13877194801

ISBN-13: 978-1-7354491-7-3 print edition
ISBN-13: 978-1-7354491-8-0 e-book edition

This book was typeset in Adobe Source Sans/Serif.
Layout: Jana Rogge

To send correspondence: aapsglobal@gmail.com

All matter originates and exists only by virtue of a force...
We must assume behind this force the existence of a
conscious and intelligent Mind.
This Mind is the matrix of all matter.

Max Planck

TABLE OF CONTENTS

Academy for the Advancement of Postmaterialist Sciences

AAPS

ACADEMY FOR THE ADVANCEMENT
OF POSTMATERIALIST SCIENCES

The Academy for the Advancement of Postmaterialist Sciences (www.AAPS-global.com) is a non-profit membership and education organization whose mission is to promote open-minded, rigorous and evidence-based enquiry into postmaterialist consciousness research. Our vision is to inspire scientists to investigate mind and consciousness as core elements of reality.

To achieve this paradigm changing mission, AAPS embraces the following values:

- Support rigorous applications of the scientific method
- Nurture curiosity and creativity in research
- Encourage open-minded exploratory and confirmatory investigations
- Model integrity and honesty in communication and education
- Value experimental and empirical data over dogma
- Create safe settings for sharing theories, evidence, and experiences
- Promote evidence-based innovation and positive societal change
- Expand awareness of the interconnectedness of all things
- Share postmaterialist evidence and understanding with the public

With these values in mind, AAPS is publishing an Advances in Postmaterialist Sciences book series to educate scientists, students, and science-minded readers about postmaterialist consciousness research and its applications. Our intent is that each volume combines rigor and creativity, expresses first person (inner experiences) as well as third person (external observations), and facilitates the betterment of humanity and the planet. Some volumes will address specific topics or themes, others will be wide ranging and diverse collections of research topics. Collectively they will help define and advance the evolution of postmaterialist theory, research and applications.

Foreword

Synchronicity and Its Implications

RICHARD TARNAS, PHD

Richard Tarnas is Professor Emeritus at the California Institute of Integral Studies, where he founded the graduate program in Philosophy, Cosmology, and Consciousness. He has taught courses in the history of ideas, depth psychology, archetypal cosmology, cultural history, and the evolution of consciousness. He has also frequently lectured on archetypal studies and depth psychology at Pacifica Graduate Institute, and was formerly the director of programs and education at Esalen Institute. He is the author of *The Passion of the Western Mind,* a history of Western thought widely used in universities, and *Cosmos and Psyche,* which received the Book of the Year Prize from the Scientific and Medical Network and contains an exploration of the phenomenon of synchronicity.

Most of us in the course of life have observed coincidences in which two or more independent events, having no apparent causal connection, nevertheless seem to form a meaningful pattern. On occasion, this patterning can strike one as so extraordinary that it is difficult to believe the coincidence has been produced by chance alone. The events give the distinct impression of having been precisely arranged, invisibly orchestrated.

Jung first described the remarkable phenomenon he named synchronicity in a seminar as early as 1928, but he continued his investigations for more than twenty years before at last attempting a full formulation in the early 1950s. He presented his influential, still-evolving analysis of synchronicity in the final paper he gave at the Eranos conferences, immediately followed by a long monograph. Developed in part through discussions with physicists, particularly Einstein and Wolfgang Pauli, the principle of synchronicity bore parallels to certain discoveries in relativity theory and quantum mechanics. Yet because of its psychological dimension, Jung's concept possessed a special relevance for the schism in the modern world view between the meaning-seeking human subject and the meaning-voided objective world. From the beginning, it has held a unique position in contemporary discussions, having simultaneously been described by physicists as posing a major challenge to the philosophical foundations of modern science, and by religious scholars as holding deep implications for the modern psychology of religion. With each decade, increasing numbers of books and heightened attention, both scholarly and popular, have been devoted to this phenomenon.[1]

Jung took particular interest in meaningful coincidences, in the beginning no doubt because their frequent occurrence had exerted a considerable impact on his own life experience. He also observed that in the therapeutic process of his patients such events repeatedly played a role, sometimes a powerful one, especially during periods of crisis and transformation. The dramatic coincidence of meaning between an inner state and a simultaneous external event seemed to bring forth in the individual a healing movement toward psychological wholeness, mediated by the integration of inner and outer realities. Such events often engendered a new sense of personal orientation within a world now seen as capable of embodying purposes and meanings beyond the mere projections of human subjectivity. The random chaos of life suddenly appeared to veil a deeper order. A subtle sign, as it were, had been given—an unexpected color in the pale void of meaning; an intimation, in William James's phrase, of "something more."

Accompanying the more profound occurrences of synchronicity was a dawning intuition, sometimes described as having the character of a spiritual awakening, that the individual was herself or himself not only embedded in a larger ground of meaning and purpose but also in some sense a focus of it. This discovery, often emerging after a sustained period of personal darkness or spiritual crisis, tended to bring with it an opening to new existential potentialities and responsibilities. Both because of this felt personal import and because of its startling metaphysical implications, such a synchronicity carried a certain numinosity, a dynamic spiritual charge with transformative consequences for the person experiencing it. In this respect, the phenomenon seemed to function, in religious terms, as something like an intervention of grace. Jung noted that such synchronicities were often kept secret or carefully guarded, to avoid the possibility of ridicule concerning an event possessing such significant personal meaning.

The classic example of a pivotal synchronistic experience is Jung's well-known description in 1951 of the "golden scarab" case:

> My example concerns a young woman patient who, in spite of efforts made on both sides, proved to be psychologically inaccessible. The difficulty lay in the fact that she always knew better about everything. Her excellent education had provided her with a weapon ideally suited to this purpose, namely a highly polished Cartesian rationalism with an impeccably "geometrical" idea of reality [as in Descartes's characteristic mode of logical demonstration]. After several fruitless attempts to sweeten her rationalism with a somewhat more human understanding, I had to confine myself to the hope that something unexpected and irrational would turn up, something that would burst the intellectual retort into which she had sealed herself. Well, I was sitting opposite her one day, with my back to the window, listening to her flow of rhetoric. She had had an impressive dream the night before, in which someone had given her a golden scarab—a costly piece of jewelry. While she was still telling me this dream, I heard something behind me gently tapping on the window. I turned round and saw that it was a fairly large flying insect that was knocking against the window-pane from outside in the obvious effort to get into the dark room. This seemed to me very strange. I opened the window immediately and caught the insect in the air as it flew in. It was a scarabaeid beetle, or common rose-chafer *(Cetonia aurata),* whose gold-green colour most nearly resembles that of a golden scarab. I handed the beetle to my patient with the words, "Here is your scarab." This experience punctured the desired hole in her rationalism and broke the ice of her intellectual resistance. The treatment could now be continued with satisfactory results. (Jung, 1951/1969, pp. 525–526)

The acute coincidence between the symbolically resonant image that the woman had experienced in her dream the night before and was just then recounting, and the spontaneous appearance at the window of an insect that was "the nearest analogy to a golden scarab that one finds in our latitudes," effectively broke through the intellectual armoring that had been blocking her psychological development. Now "her natural being could burst through ... and the process of transformation could at last begin to move." (Jung, 1952/1969, pp. 438–439).

In another such instance, recounted in Esther Harding's notebooks, a patient whose dreams were filled with sexual imagery kept attempting to interpret the dreams in non-sexual symbolic terms, despite Jung's efforts to persuade her of their more plausible straightforward meaning. On the day of her next appointment, two sparrows fluttered to the ground at the woman's feet and "performed the act" (Harding, 1977, pp. 182–83).

> On rare occasions a synchronicity proves to have an extraordinary power, through its impact on an historically significant individual, so that it ultimately plays a pivotal role in the collective life of the larger culture. The famous coincidence that formed a turning point in the life of Petrarch took place at the climax of his ascent of Mont Ventoux in April 1336, an event that has long been regarded by scholars as representing the symbolic beginning of the Renaissance. For many years Petrarch had been sensing a growing impulse to ascend the mountain, to see the vast panorama from its peak, though doing such a thing was virtually unheard of in his time. Finally choosing the day, with his brother for a companion, he made the long ascent, marked by intense physical exertion and inward reflection. When he at last attained the summit, with clouds below his feet and winds in his face, Petrarch found himself overwhelmed by the great sweep of the world that now opened out to him—snow-capped mountains and the sea in the distance, rivers and valleys below, the wide expanse of skies in every direction. James Hillman recounts the event:
>
>> At the top of the mountain, with the exhilarating view of French Provence, the Alps, and the Mediterranean spread before him, he had opened his tiny pocket copy of Augustine's *Confessions*. Turning at random to book X, 8, he read: "And men go abroad to admire the heights of mountains, the mighty billows of the sea, the broad tide of rivers, the compass of the ocean, and the circuits of the stars, and pass themselves by ..."
>>
>> Petrarch was stunned at the coincidence between Augustine's words and the time and place they were read. His emotion both announced the revelation of his personal vocation and heralded the new attitude of the Renaissance. ... Petrarch draws this crucial conclusion from the Mont Ventoux event: "Nothing is admirable but the soul" *(nihil praeter animum esse mirabile).* (Hillman, 1975, pp. 195–196)

Petrarch was so moved by the coincidental force of the passage that he remained silent for the entire descent down the mountain. He at once recognized the coincidence as part of a larger pattern of such transformative moments which had happened to others in the history of spiritual conversions: "I could not believe that it was by a mere accident that I happened upon them. What I had there read I believed to be addressed to me and to no other, remembering that St. Augustine had once suspected the same thing in his own case" (Petrarch, 1336/1898, p. 317). For indeed Augustine had undergone a nearly identical experience at his own momentous spiritual turning point: In the garden of Milan in 386, in a frenzy of spiritual crisis, he heard a child's voice from a nearby house mysteriously repeating the words "Tolle, lege" ("Pick up and read"). Uncertain of their significance, he finally opened at random a copy of St. Paul's epistles and there read words that spoke with uncanny precision to the nature of his lifelong conflict and its resolution, immediately after reading which "the light of certainty flooded my heart and all dark shadows of doubt fled away" (*Confessions,* VIII, 29).

Here too, Augustine's emotion in the garden of Milan both announced the revelation of his personal vocation and heralded the new attitude of the Christian epoch being born. A thousand years later, it would be Augustine's own words randomly encountered that provided a strikingly similar catalyzing force for Petrarch on Mont Ventoux. This time the synchronistic epiphany would unfold in a new direction and with different consequences—one revelation in the garden, pointing to Christianity and the Middle Ages, the other on the mountain, pointing to the Renaissance and modernity.

***

Jung believed that synchronicities generally seemed to serve the same role as dreams, psychological symptoms, and other manifestations of the unconscious, namely, to compensate the conscious attitude and move the psyche from a problematic one-sidedness toward greater wholeness and individuation. Not only did the unexpectedly externalized pattern of meaning seem to represent more than mere chance coincidence, it also appeared to serve a definite purpose, impelling the psyche toward a more complete psychological and spiritual realization of the individual personality. This self-realization was achieved through a deeper integration of conscious and unconscious, which ultimately required of the individual a discerning surrender of the usual conscious attitude of knowing superiority. In this view, the perceptive interpretation of synchronistic phenomena, as with all expressions of the unconscious, rather than inflating the egocentric importance of the individual in a narcissistic manner, could correct precisely these tendencies and open the psyche to a larger vision.

An instructive example of this self-critical, compensatory approach toward synchronicity in Jung's own life was recounted by Henry Fierz, describing a meeting with Jung in the 1950s. Fierz had come to discuss Jung's assessment concerning the possible publication of a manuscript by a scientist who had recently died. At the appointed hour of five o'clock, Fierz arrived for the meeting and the discussion began:

> Jung had read the book and he thought that it should not be published, but I disagreed and was for publication. Our discussion finally got rather sharp, and Jung looked at his wristwatch, obviously thinking that he had spent enough time on the matter and that he could send me home. Looking at his watch he said: "When did you come?" I: "At five, as agreed." Jung: "But that's queer. My watch came back from the watchmaker this morning after a complete revision, and now I have 5:05. But you must have been here much longer. What time do you have?" I: "It's 5:35." Whereon Jung said: "So you have the right time, and I the wrong one. Let us discuss the thing again." This time I could convince Jung that the book should be published. (Fierz, 1982, p. 21)

Here the synchronistic event is of interest not because of its intrinsic coincidental force but because of the meaning Jung drew from it, essentially using it as a basis for challenging and redirecting his own conscious attitude. The unexpected stopping and resulting error of the watch was immediately recognized by Jung as paralleling—and as thereby bringing to his attention—what he now suspected might be a comparable stoppage and error in his own thinking about the matter at hand. He was alert to the fact that the two events, one inner and the other outer, would have taken place at virtually the same moment. Rather than automatically assuming that there could be no significant connection between the state of his watch and the state of his thinking, which would certainly be the usual assumption, his immediate intuition was of a larger field of meaning underlying and patterning all that happened in the room at the time. In that field, events having no apparent causal connection in the conventional sense could be recognized as participating in a more subtly ordered whole, a larger pattern of meaning that was discernible to the prepared mind—even if that meaning challenged his conscious attitude.

For Jung, the symbolic connection between the two events was as transparently intelligible as if he were reading a newspaper, and he acted accordingly. What made the correlation between the inner and outer events intelligible was the presence of two factors: first, a developed capacity for thinking and perceiving symbolically, a cultivated sensitivity to metaphoric and analogical patterns that connect and thereby illuminate diverse phenomena; and second, an epistemological openness to the possibility that such meaning can be carried by the outer world as well as the inner, by all of nature and one's surrounding environment and not just by the human psyche.

Yet the recognition of synchronicities involves subtle judgments made in circumstances usually pervaded by ambiguity and open to multiple interpretations. The suggestive patterning and often delicate precision of detail in such coincidences notoriously escapes the net of objectivistic assessments and experimental tests. Synchronicities seem to constitute a lived reality the experience of which depends deeply on the sensitive perception of context and nuance. For synchronicities have a shadow side, as in the exaggeration of the trivial to discover a self-inflating meaning. Similarly, the paranoid's morbidly narrow interpretation of coincidences in terms of other people's malign plots cunningly directed at the self, or psychotic delusions of self-reference. Such interpretations are, as Jung once suggested, pre-Copernican, egocentric. They center the world of meaning naively on the old narrow self, inflating the separate ego, or persecuting it, and thereby evade the more complex and painful emergence of the individuated self that is in dialogue with the whole.

Such an emergence requires attending to the claims and communications of the larger cosmos of the unconscious. A painstaking cultivation of self-knowledge must be undertaken to avoid succumbing to mere projection. Discriminating such events requires a self-critical awareness of unconscious tendencies towards narcissistic distortion, by which random or peripheral events are continually transformed into signs of an egocentric universe. No less crucial is the development and balanced interplay of multiple faculties of cognition—empirical, rational, emotional, relational, intuitive, symbolic. A capacity for acute yet balanced discernment has to be forged, founded not only on an alertness to meaning-ful pattern but also on a disciplined mindfulness of the larger whole within which the individual self seeks orientation.

Today, a half century after Jung's original formulation of the principle of synchronicity, with both the concept and the phenomenon now so widely recognized, one can discern a typical sequence and progression in the nature of synchronistic events and responses. The first stage is usually marked by the experience of various ambiguously suggestive coincidences and patternings which may seem somewhat remarkable, curious, or even vaguely uncanny, but can still be regarded as perhaps merely fortuitous or subjective, and are therefore more or less ignored and forgotten. Eventually, there may occur one or more especially powerful synchronicities, unambiguous in their coincidental force and precision of patterning, which have a revelatory impact on the individual and mark a decisive threshold in his or her psychological and spiritual development. Not infrequently, synchronicities of this category occur in association with births, deaths, crises, and other major turning points in life. On occasion, there may take place a sudden convergence of many such synchronicities, intricately interconnected, occurring in close proximity or in rapid succession, and having the effect of an overpowering epiphany of new meaning and purpose in the life of the individual.

Over time, however, after this threshold has been crossed, a new attitude toward synchronicities often emerges as their frequency and character come to seem part of life's pervasive intelligence and artistry—less a paradigm-shifting revelation of a new order of reality and more a continuing source of meaning and orientation with which to participate in life with greater sensitivity and intelligence. A disciplined alertness to significant pattern in the outer world as well as inner begins to develop as an essential aspect of living a more conscious life. The occurrence of synchronicities is seen as permitting a continuing dialogue with the unconscious and the larger whole of life, while also calling forth an aesthetic and spiritual appreciation of life's powers of symbolically resonant complex patterning.

Although Jung himself did not explicitly describe this later stage in his principal monograph on synchronicity, it is evident from many scattered passages in his writings as well as from the recollections and memoirs of others that he both lived his life and conducted his clinical practice in a manner that involved a constant attention to potentially meaningful synchronistic events which would then shape his understanding and actions. Jung saw nature and one's surrounding environment as a living matrix of potential synchronistic meaning that could illuminate the human sphere. He attended to sudden or unusual movements or appearances of animals, flocks of birds, the wind, storms, the suddenly louder lapping of the lake outside the window of his consulting room, and similar phenomena as possessing possible symbolic relevance for the parallel unfolding of interior psychological realities. For the woman who had the dream of the golden scarab, the next day's synchronistic visitation through the window was dramatically transformative, whereas the same event for Jung represented a striking but not uncharacteristic example of the meaningful patterning of inner and outer events to which he had long learned to be attentive.

In sharp contrast to the modern world view, Jung ceased to regard the outer world as merely a neutral background against which the human psyche pursued its isolated intrasubjective quest for meaning and purpose. Rather, all events, inner and outer, whether emanating from the human unconscious or from the larger matrix of the world, were recognized as sources of potential psychological and spiritual significance. From this perspective, not only the individual psyche, and not only humanity's collective unconscious, but all of nature supported and moved the human psyche towards a larger consciousness of purpose and meaning.[2] Each moment in time possessed a certain tangible character or quality which pervaded the various events taking place at that moment.

> It seems, indeed, as though time, far from being an abstraction, is a concrete continuum which contains qualities or fundamentals which can manifest themselves in relative simultaneousness in different places and in a parallelism which cannot be explained, as in cases of simultaneous appearance of identical thoughts, symbols, or psychic conditions.... Whatever is born or done at this particular moment of time has the quality of this moment of time."[3] (Jung, 1930/1966, p. 56)

Central to Jung's understanding of such phenomena was his observation that the underlying meaning or formal factor that linked the synchronistic inner and outer events—the formal cause, in Aristotelian terms—was archetypal in nature. Building on insights from Freud, and drawing from the classical Platonic philosophical vocabulary as well as from esoteric traditions, Jung had long regarded and defined the archetypes as the fundamental governing principles of the human psyche. On the basis of his own analyses as well as that of others, not only of a diverse range of clinical phenomena but also of the art, myths, and religions of many eras and cultures, Jung had come to view archetypes as innate symbolic forms and psychological dispositions that unconsciously structure and impel human behavior and experience at both the personal and collective levels. They are "self-portraits" of the instincts and render human experience meaningful according to certain timeless universal patterns or forms—Light and Dark, Birth and Death, Rebirth, the Hero, the Great Mother, the Child, the Trickster, the Shadow, Good and Evil, Eros and Logos, Feminine and Masculine, as well as more specifically personified and culturally inflected forms such as Aphrodite, Oedipus, Dionysus, Prometheus, Saturn, Shakti, Kali, Shiva, Wotan, Isis, and Sophia. Another major category of the archetypes comprised the mathematical principles of number and geometric form, as in the Pythagorean-Platonic tradition, as well as traditional sacred forms such as the mandala, the circle, and the cross. All these principles were seen as possessing a primordial, mythic, numinous character grounded in the deepest layers of the psyche, expressing a collective unconscious shared by all human beings.

During most of his career, Jung worked and wrote within the modern Cartesian-Kantian philosophical framework of a basic division between the human subject and the objective world, and thus tended to restrict archetypes to the interior world of the human psyche. His view of archetypes in the early and middle periods of his career was generally equivalent to Kant's notion of a priori forms and categories: They were inherited psychological structures or dispositions that preceded and shaped the character of human experience but could not be said to transcend the human psyche. In his later work, however, and most explicitly in the context of his analysis of synchronicities, Jung moved towards a conception of archetypes as autonomous patterns of meaning that inform both psyche and matter, providing a bridge between inner and outer: "Synchronicity postulates a meaning which is a priori in relation to human consciousness and apparently exists outside man" (Jung, 1952/1969, p. 501–502). Jung's later work thus intimated the ancient understanding of an ensouled world, of an *anima mundi* in which the human psyche participates and with which it shares the same ordering principles of meaning. Jung noted parallels between synchronistic phenomena and the Chinese understanding of the Tao, the ancient Greek conception of the cosmic sympathy of all things, the Hermetic doctrine of microcosm and macrocosm, the medieval and Renaissance theory of correspondences, and the medieval concept of the preexistent ultimate unity of all existence, the *unus mundus* (the unitary world).[4]

In each case of synchronicity, Jung discerned an underlying archetypal coherence that linked the otherwise unconnected events, informed the larger field of meaning, and gave to the time of the synchronicity's occurrence a specific fundamental quality. In the first case cited above, for example, the symbolically charged image of the golden scarab expressed the archetypal principle of rebirth and renewal, visible in the Egyptian myth of the Sun-god who in the netherworld during the night sea journey changes himself into a scarab, then mounts the barge to rise again reborn into the morning sky at dawn. In Egyptian religion, the mythic journey of the Sun mediated the spiritual journey of the soul, providing the individual with a transformative symbolic pattern of descent and renewal, death and rebirth.

The case of the stopped watch, by contrast, was pervasively informed by the complex archetype of Saturn-Kronos, the senex principle, a central symbol and figure in the Western cultural tradition from the ancient Greek and Hellenistic era through the Middle Ages and Renaissance.[5] In this synchronicity, the Saturn archetype was visible not only in all the concrete details involving time, but also in the intricately interrelated themes of stoppage and being stuck (in both mind and watch), of opposition and rejection, error, fault, correction, judgment and self-judgment, the superego. Each element and stage of the event suggested another dimension of the Saturn principle's multivalent spectrum of meanings: the precise meeting time, the task at hand, the problem to be solved, the pronouncement of judgments, the strife of disagreement, the attempt to bring an end to the task, the careful checking and comparing of the time, the act of negation and criticism first directed outward towards the other and then inward towards the self, the self-correction followed by repetition, engaging the problem again and trying this time to get it right. Finally, the overarching themes, deciding the fate of the manuscript, judging the legacy of the deceased, death as the stopping of time: all characteristic expressions of Saturn and the senex discernible within the hour of the meeting.

Because synchronicities seemed to reflect and embody the same archetypal forms that Jung and many others had come to see as basic underlying principles of the human psyche, the occurrence and recognition of such meaningful coincidences gave a crucial new dimension to the archetypal perspective. The empirical conformity between the event occurring in the external world and the archetypal quality of the internal state of consciousness suggested that the active archetype could not be localized as an exclusively subjective intrapsychic reality. Rather, both psyche and world, inner and outer, were informed by the archetypal pattern and thereby united by the correlation. It was specifically the experiential potency of this spontaneous archetypal resonance that seemed to act as a healing solvent on the hardened polarities—between self and world, subject and object, conscious and unconscious—of the person experiencing the synchronicity.

> The collective unconscious surrounds us on all sides. ... It is more like an atmosphere in which we live than something that is found *in* us. ... Also, it does not by any means behave merely psychologically; in the cases of so-called synchronicity it proves to be a universal substrate present in the environment rather than a psychological premise. Wherever we come into contact with an archetype we enter into relationship with transconscious, metapsychic factors. (Jung, 1946/1973, p. 433)

This development in Jung's thought thus constituted a major shift in his understanding of the religious situation confronting the modern psyche. From early in his career, Jung saw both the psychological and the spiritual path of the modern self as requiring a sustained direct encounter with the archetypal unconscious. Here lay the possibility not only of deeper psychological self-awareness but also of spiritual transformation, permitting an engagement with those numinous realities that could profoundly heal the psyche and provide it with an orienting purpose and transcendent meaning. Throughout most of his writings, this engagement was understood as taking place within what Jung essentially regarded as the sacred circle of the human psyche. Eventually, however, Jung's many years of studying synchronicities moved him to recognize this engagement as something that is enacted within the larger sacred circle of nature as a whole. In this perspective, not just the interior depths of the human psyche but the interior depths of nature itself supported the unfolding of human spirituality and each person's struggle towards individuation.[6]

The recognition of synchronicities' potential metaphysical implications not only suggested a transformation in the psychology of religion, it represented a critical step towards bridging the schism between religion and science in the modern era, so long embodied in the seemingly unbridgeable chasm between psyche and world. As the physicist Victor Mansfield has written, speaking for many: "I have encountered too many synchronistic experiences, both in my life and that of others, to ignore them. Yet these surprisingly common experiences pose tremendous psychological and philosophical challenges for our world view. They are especially troubling experiences for me as a physicist trained within the culture of scientific materialism" (Mansfield, 1995, p. 1). With these implications in mind, both philosophical and psychological, Jung's student and close associate Marie-Louise von Franz stated in an interview late in her life that "the work which has now to be done is to work out the concept of synchronicity. I don't know the people who will continue it. They must exist, but I don't know where they are" (von Franz, 1990/1995, p. 22).

Despite synchronicities' enigmatic and often readily dismissed character, it was with such humble clues that Jung began to open up the possibility of a fundamental redefinition of both the modern religious situation and the scientific world picture, beyond the closed universe of the spiritually aspiring psyche encompassed by a disenchanted world. The existence of synchronicities implied that the world could no longer be seen as a definitively meaningless void. The dynamic relationship between different dimensions of being—both between the human self and the encompassing world, and between consciousness and the unconscious—had to be reconceived. It appears to have been Jung's growing recognition of the magnitude of these implications for the modern world view that impelled him to labor so strenuously, even courageously, to bring critical awareness of the phenomenon of synchronicity into the intellectual discourse of the twentieth century.[7]

The psychological and spiritual quest of the modern self now extended beyond an exclusively subjective, intrapsychic horizon, for that quest took place within the matrix of a world that evidently possessed an intrinsic capacity for expressing and supporting meaning and purpose. Subtly and tenuously, but unmistakably, the larger context within which the modern psyche pursued its search for wholeness had begun to shift.

Of course it is only in the modern era that synchronicity becomes an issue, something puzzling, disturbing, or amazing. For most of human history, there was no need to single out and name a particular category of anomalous events to describe what we call synchronicities. For primal tribal and contemporary indigenous societies, such events were and are simply an expression of the ensouled, animate field of all reality, the anima mundi. In archaic civilizations, what we call synchronicities represented signs and omens from the gods in a cosmos in which above and below, the celestial and terrestrial, were experienced as embodying a deeper meaningful unity. And in societies informed by one of the Axial religions such as Judaism, Christianity, or Islam, or one of the Axial philosophical traditions like Platonism or Taoism, synchronicities would simply be seen as an expression of Providence, of God's will, of the Tao, of the Universal Mind's ordering of reality.

For the modern mind, however, synchronicities constitute an intrinsically controversial phenomenon, as they seem to undermine the global assumption that we live in a randomly evolving universe without intrinsic meaning or purpose. It is only in the disenchanted modern world view, where the buffered rational subject regards all meaning and purpose in the cosmos as originating in the human being, that meaningful coincidences become something to grapple with or dismiss, depending on the degree of epistemic armoring. It is thus perhaps for the buffered modern sensibility that synchronicities can carry the most profound metaphysical impact, be the most radically transformational, precisely because they point towards a revisioning of not only our overarching cosmology but our underlying identity.

As a distinguished scholar and professor of religious studies once confided to me, "I have stopped believing in most things, but I do believe in synchronicity."[8] So it is that in our postmodern age, marked by a disorienting radical pluralism of perspectives and a pervasive indeterminacy about the ultimate nature of reality, we find ourselves facing synchronistic phenomena with new eyes. The postmodern is itself a crucible of transformation. Our civilization has clearly reached a threshold in human history, and indeed synchronicities seem to occur in individual lives with greater frequency and intensity during major thresholds of human life, births, deaths, crises, sudden openings. Perhaps what is true for individuals is true as well for a civilization.

It is within this newly expansive, open-ended, and metaphysically suggestive context that the scholars, scientists, and researchers who have contributed to the present volume share with us representative synchronicities and explore their larger implications. In so doing, they seek to engage a mystery that may hold considerable significance for humanity's quest to understand itself and its relationship to a world that holds more things than are dreamt of in our conventional modern philosophies.

***

An earlier version of this essay appeared in Richard Tarnas, *Cosmos and Psyche: Intimations of a New World View* (New York: Viking, 2006), 50-60, 494–500.

Notes

1. The primary texts by Jung are the 1951 Eranos conference lecture "On Synchronicity" and the longer 1952 monograph *Synchronicity: An Acausal Connecting Principle,* both in *Collected Works of C. G. Jung,* trans. R. F. C. Hull, ed. H. Read, M. Fordham, G. Adler, W. McGuire, Bollingen Series XX (Princeton: Princeton University Press, 1953–79), vol. 8. While invaluable and seminal, however, these works by Jung are marked by many conceptual incoherencies and confusions, perhaps inevitable for the first presentation of a new principle of understanding so radically challenging of existing assumptions. Moreover, as discussed in note 5 below, Jung's lecture and monograph focused on categories of phenomena such as paranormal events and experimental data from physics which obscured the human, psychologically transformative dimension of synchronistic phenomena, though the latter was in fact far more central to Jung's own experience and clinical observations.

   Other relevant texts by Jung can be found in *Jung on Synchronicity and the Paranormal,* edited and with an introduction by Roderick Main (Princeton: Princeton University Press, 1997). For synchronicity's implications for the psychology of religion, see Robert Aziz, *C. G. Jung's Psychology of Religion and Synchronicity* (Albany: State University of New York Press, 1990): "The synchronicity concept is, arguably, the single theory with the most far-reaching implications for Jung's psychology as a whole, particularly for his psychology of religion" (p. 1). For physics and the scientific world view, see Victor Mansfield, *Synchronicity, Science, and Soul-Making* (Chicago: Open Court, 1995). Other significant writings on synchronicity include books or essays by Arthur Koestler, Antony Flew, Michael Fordham, Ira Progoff, Marie-Louise von Franz, Aniela Jaffé, Allan Combs, Mark Holland, Michael Conforti, Jean Shinoda Bolen, David Peat, Sean Kelly, and Ray Grasse.

2. Robert Aziz comments: "For Jung, the call to individuate arises from the deepest sources of life and is supported inwardly and outwardly by the compensatory activities of nature. It is a call, therefore, that is not to be taken lightly. Both inwardly and outwardly nature strives unceasingly to bring about the realization, in the life of the individual, of a unique pattern of meaning. ... [A]s evidenced by his own writings on synchronicity and, perhaps more importantly, by the way he lived his own life, the individuation process extends beyond the psychologi-

cal realm and assumes the character of a drama that takes the whole of nature for its stage. What we normally regard as the discontinuous inner and outer worlds become enclosed within the same circle of wholeness. Inwardly and outwardly nature works, through the compensatory patterning of events, to further the movement of the individual toward wholeness. ... Now one is challenged to achieve a full understanding of the meaning that conjoins one, not only to the unconscious, but to nature in its entirety. This is the new spiritual challenge of individuation. It is the task of experiencing within the sacred circle of nature as a whole the meaning of an individual existence" (*C. G. Jung's Psychology of Religion and Synchronicity,* pp. 165–166).

Jung's attitude closely resembles the characteristic primal and shamanic alertness to nature's symbolically significant patterning, as well as ancient Chinese Taoist philosophy, in which the dominant principles are pattern, order, symbolic correlations, the unity of human and cosmos, and the interdependence of all things. Cf. Richard Wilhelm, *Secret of the Golden Flower:* "Chinese philosophy is built on the premise that the cosmos and man, in the last analysis, obey the same law; that man is a microcosm and is not separated from the macrocosm by any fixed barriers. The very same laws rule for the one as for the other, and from the one a way leads into the other. The psyche and the cosmos are to each other like the inner world and the outer world. Therefore man participates by nature in all cosmic events, and is inwardly as well as outwardly interwoven with them. (p.11, quoted in Aziz, op. cit., p. 135.) Cf. also Joseph Needham, *Science and Civilisation in China:* "The key-word in Chinese thought is *Order* and above all *Pattern.* ... The symbolic correlations or correspondences all formed part of one colossal pattern. Things behaved in particular ways not necessarily because of prior actions or impulsions of other things, but because their position in the ever-moving cyclical universe was such that they were endowed with intrinsic natures which made that behaviour inevitable for them. ... They were thus parts in existential dependence upon the whole world-organism" (vol. 2, p. 281, quoted in Aziz, op. cit., pp. 135–36).

3. This formulation of synchronicity in terms of *qualitative time,* from 1930, reflects Jung's early astrological research, beginning in 1911, and his experiments throughout the 1920s with the *I Ching:* "Jung's first theorizing about synchronicity was done with reference to astrology and the I Ching and focused on the fact that things arising in a particular moment of time all share the characteristics of that moment. It appears to have been this understanding of the role of time, an

understanding in which simultaneity does indeed play an essential part, which led Jung to coin the term 'synchronicity,' with its emphasis on the element of time (Gk. *syn* = together, *chronos* = time)" (R. Main, *Jung on Synchronicity and the Paranormal,* p. 23).

In later years, Jung increasingly focused on synchronicity's parallels with twentieth-century physics, influenced first by his conversations about relativity with Einstein (a dinner guest on several occasions in the 1909-12 period), and then several decades later through his discussions about quantum mechanics with his patient and friend Wolfgang Pauli. Reflecting the parallels with both physics and the parapsychological experiments of J. B. Rhine at Duke University that had started in the 1930s, Jung began to broaden the concept of synchronicity to include many phenomena—various paranormal phenomena such as precognition and telepathy, the discontinuities of modern physics, the properties of natural numbers—for which simultaneity, qualitative time, and meaning were not always relevant factors. Instead, he began to stress "the psychic relativization of time and space" and "general acausal orderedness." However, as many commentators have noted (Koestler, Aziz, Mansfield, Main), some of these additions stretched the parameters of the concept to include phenomena for which the original term "synchronicity" was now problematic and less obviously appropriate.

In his efforts to include the various phenomena from parapsychology and physics, Jung was essentially combining into one overarching concept several separate classes of events which in many cases seemed to confuse and overlook their fundamental differences. For example, an especially important difference was concealed by Jung's conflation of two basic categories in his discussions, the meaningful coincidence of simultaneous events and the experience of clairvoyant cognition. The first category could be said to represent the classic form of synchronicity, illustrated by the paradigm case of the golden scarab, in which the outer world unexpectedly brought forth a concrete external event that closely paralleled in meaning an inner psychological state. The second category was centered on cases in which an individual experienced internally—by intuition, dream, or vision—some external event in a future time or at a distant location.

But it is only in the first category—the one upon which Jung placed far greater emphasis—that the crucial possibility of a meaning-embedded world presents itself, with all the metaphysical implications associated with the concept of synchronicity. By contrast, in cases involving paranormal experiences like clairvoyance, telepathy, and precognition, the individual in question can simply be seen as exercising some not yet understood perceptual or cognitive faculty, with no implications for the outer world's capacity to embody meaning in a manner that transcends the human psyche. Such cases would thus only provide evidence pointing to the need for a revised understanding of human abilities and the parameters of human consciousness. Other cases such as those involved in the parapsychological experiments and in quantum physics present comparable differences, as well as additional ones that eliminate the presence of meaning altogether as a central factor.

It seems likely that the scientific status of physics and the statistical-experimental nature of the parapsychological research played a significant role in moving Jung in the above direction, encouraging him to modify the terms of his original concept and add further categories to it, in his hope to elevate the viability of his challenging hypothesis within the science-dominated intellectual ethos of the mid-twentieth century (see, for example, Ira Progoff, *Jung, Synchronicity, and Human Destiny* [New York: Dell, 1973], p. 143; Aziz, op. cit., p. 2; Mansfield, op. cit., pp. 33–34; and Main, op. cit., pp. 15–17, 23–27.) This also perhaps explains Jung's 1954 letter to André Barbault where he contrasted synchronicity with the qualitative time hypothesis he had earlier proposed, suggesting now that the idea of synchronicity replaced rather than manifested the principle of qualitative time (C. G. Jung, *Letters 2: 1951–1961,* ed. G. Adler, A. Jaffé, trans. R. F. C. Hull [London: Routledge & Kegan Paul, 1976], p. 176). Jung here described the latter concept as tautological because "time in itself consists of nothing," and is only "qualified" or defined by events.

This apparent change in Jung's position probably reflected not only his larger shift in emphasis towards a more broadly inclusive and science-oriented formulation of the synchronicity principle, but also his desire to locate the essence of synchronistic phenomena in the parallel patterning of the phenomena themselves rather than in some a priori characteristics of time apart from the phenomena. Thus, contrary to some commentaries on this topic, Jung's later objection was not to the idea of time having a de facto qualitative dimension, but rather to the idea that time *itself* was the a priori determining factor of the observed qualities. Instead, Jung clearly viewed the determining factor, in the sense of what "arranged" the qualitative patterning within the flux of events, to be not time per se, but rather the constellated archetype. In Aristotelian terms, the archetype is the *formal cause* of the synchronicity. Time is thus regarded as possessing a qualitative dimension, but the quality that will be manifest at any given time is indeterminate and potential until a specific archetype is constellated.

In retrospect, by the time Jung wrote his principal analysis on synchronicity in the early 1950s, his increasing focus on parapsychology and physics had, in a sense, partly colonized the original concept, and thus obscured the reality of how he had integrated his experience of meaningful coincidences into his life and clinical practice. Jung's original and most familiar examples of synchronicities emerged in psychological, therapeutic, religious, divinatory, and esoteric contexts (as compared with experi-mentally tested telepathy, precognition, clairvoyance, psychokinesis, out-of-body and near-death experiences and other paranormal phenomena, as well as the discontinuities of physics, for all of which alternative explanations have been developed that are arguably more apt than synchronicity). For the more characteristic original categories of meaningful coincidences, such as the golden scarab and stopped watch examples cited in the text, the idea of qualitative time (i.e., time as possessing a qualitative dimension), along with the elements of simultaneity and meaning, clearly remains relevant as part of a larger theoretical conception in which the archetypal meaning informing and connecting the synchronistic events serves as a fundamental explanatory principle.

Many aspects of relativity and quantum mechanics are indeed relevant to synchronistic phenomena—a relativized space-time continuum, the collapse of strict linear causality and a fully independent objective world, complementarity, probabilistic indeterminacy, nonlocality, the interconnected and interdependent nature of reality. Other essential elements of synchronicities, however, have no parallels in physics—above all, the fundamental presence of meaning as the structuring factor, and the apparent teleological or purposive aspect of such events.

In traditional philosophical terms, these two basic elements of synchronicities—meaning and purpose—represent straightforward expressions of what Aristotle called *formal and final causes,* respectively. Compared with the simpler (or simplistic) modern view of causality, which is entirely linear-mechanistic in nature, Aristotle's more nuanced and capacious formulation defined "cause" as that which is a necessary, though not in itself sufficient, condition for the existence of something. Such conditions included formal and final (teleological) factors, as well as the material and efficient factors stressed by mainstream modern science. When Jung originally employed the term "acausality" and emphasized that synchronicities were fundamentally acausal in nature, he provided a helpful and probably necessary counterpoint to the narrow conventional scientific understanding of linear-mechanistic causality that was then (the first half of the twentieth century) nearly ubiquitous. But the clear applicability of Aristotle's richer classical formulation of causality for understanding synchronicities within a Jungian archetypal and teleological perspective places in question the continued usefulness or appropriateness of the term "acausality" in this context.

4. Marie-Louise von Franz: "The most essential and certainly the most impressive thing about synchronistic occurrences, the thing which really constitutes their numinosity, is the fact that in them the duality of soul and matter seems to be eliminated. They are therefore an empirical indication of the ultimate unity of all existence, which Jung, using the terminology of medieval natural philosophy, called the unus mundus. In medieval philosophy this concept designates the potential preexistent model of creation in the mind of God in accordance with which God later produced the creation. It is, according to John Scotus Erigena, 'God's vital or seminal power which changes from a Nothing, which is beyond all existence and non-existence, into countless forms.'" M.-L. von Franz, *C. G. Jung: His Myth in Our Times* (Toronto: Inner City Books, 1998), p. 247.

5. *Senex*, from the Latin: old man, elder, age, as in senescence, senator, senility. Reflecting the complex and ambiguous constellation of qualities, positive and problematic, characteristic of old age, from narrowness, rigidity, and pessimism, and the concern with order, control, and death, to gravitas, experience, and wisdom, the *senex* is closely associated with the Saturn archetype. The senex is the polar complement to the *puer,* or *puer eternus* (eternal child), with which it is mutually implicated. For a richly perceptive exploration of the senex, see two early papers by James Hillman, "On Senex Consciousness," *Spring 1970: An Annual of Archetypal Psychology and Jungian Thought* (Dallas: Spring Publications, 1970), pp. 146–165; and his 1967 Eranos paper, "Senex and Puer: An Aspect of the Historical and Psychological Present," in J. Hillman (ed.), *Puer Papers* (Dallas: Spring Publications, 1991), pp. 3–53.

6. This understanding of Jung's evolution is at the center of Aziz's analysis: "In 1937 when Jung exhorted his Yale audience to move beyond the confines of established religion and accept the challenge of 'immediate religious experience,' what Jung had in mind was for them to enter consciously into a direct encounter with the unconscious. For those for whom the rituals of conventional religion

had lost their efficacy, what Jung offered as an alternative was an intrapsychic ritual which, properly followed, would lead to the emergence of a highly personalized spiritual wholeness. What Jung had in mind in 1937, then, was a ritual to be enacted within the sacred circle of the psyche. ... [H]owever, this earlier Jungian notion of religious ritual has been dramatically transformed by the synchronicity concept, indeed, so much so that we can now say that Jung's notion of 'immediate religious experience' may be taken to refer not simply to an intrapsychic encounter, but to a direct encounter with nature in its entirety. The Jungian ritual, to put it simply, is now a ritual which is to be enacted within the sacred circle of nature as a whole. ... [F]or Jung the 'holy' is encountered as much outwardly, in the synchronistic patterning of events, as it is inwardly. Accordingly, the individual in search of 'immediate religious experience' is now required to attend to the compen-satory images with which nature presents one outwardly with the same religious seriousness with which one attends to the 'images of wholeness offered by the unconscious.' ... The religious need, as Jung puts it, longs for wholeness, and here the wholeness to which one must open oneself is a wholeness that is not only transmitted intrapsychically, but transmitted to the individual through the synchronistic patterning of events in one's environment (Aziz, op. cit., pp. 167–168).

7. Cf. R. Main, "Religion, Science, and Synchronicity," *Harvest: Journal for Jungian Studies* 46, no. 2 (2000), pp. 89–107. "In a 1955 letter to R. F. C. Hull, Jung reported: 'The latest comment about "Synchronicity" is that it cannot be accepted because it shakes the security of our scientific foundations, as if this were not exactly the goal I am aiming at...' On the same day he wrote to Michael Fordham of 'the impact of synchronicity upon the fanatical one-sidedness of scientific philosophy.' Specifically, Jung thought that his work on synchronicity demonstrated the need to expand the current conception of science in order to include, in addition to the classical concepts of time, space, and causality, a principle of acausal connection through meaning. This, he concluded, would introduce the psychic factor of meaning into our scientific picture of the world, help get rid of 'the incommensurability between the observed and the observer', and make possible a 'whole judgement'—that is, a judgement that takes into consideration psychological as well as physical factors. Because for Jung the psychological mediates between the physical and the spiritual, to link the physical and psychological in this way entails setting up a potential bridge between the physical and the spiritual, hence between science and religion. These bold conclusions and

implications from Jung's work on synchronicity resonate with many subsequent attempts to develop more holistic models of science—some directly exploring Jung's suggestions, for instance those of David Peat and Victor Mansfield, others working independently but aware of Jung's contribution and possibly influenced or inspired by it, for instance those of David Bohm and Rupert Sheldrake."

8. Jeffrey Kripal, private communication, Esalen Institute, March 2004.

References

Fierz, H. (1982). In F. Jensen (Ed.), *C. G. Jung, Emma Jung and Toni Wolff: A collection of remembrances.* San Francisco: The Analytical Psychology Club of San Francisco.

Harding, E. (1977). In W. McGuire, & R. F. C. Hull (Eds.), *C. G. Jung Speaking: Interviews and Encounters.* Princeton, NJ: Princeton University Press.

Hillman, J. (1975). *Re-visioning psychology.* New York: Harper & Row.

Jung, C. G. (1930/1966). Richard Wilhelm: In memoriam (R. F. C. Hull Trans.). In In H. Read, M. Fordham, G. Adler, & W. McGuire (Eds.), *Collected works of C. G. Jung:* Vol. 15, *The spirit of man, art, and literature* (p. 56). Princeton, NJ: Princeton University Press. (Lecture originally delivered and published 1930.)

Jung, C. G. (1946/1973). Letter of July 10, 1946 (R. F. C. Hull, Trans.). In G. Adler & A. Jaffé (Eds.), *C. G. Jung letters volume 1: 1906–1950.* Princeton, NJ: Princeton University Press.

Jung, C. G. (1951/1969). On synchronicity (R. F. C. Hull, Trans.). In H. Read, M. Fordham, G. Adler, & W. McGuire (Eds.), *Collected works of C. G. Jung:* Vol. 8, *The structure and dynamics of the psyche* (pp. 525–526, 2nd ed., 1969). Princeton, NJ: Princeton University Press. (Lecture originally delivered 1951)

Jung, C. G. (1952/1969). Synchronicity: An acausal connecting principle (R. F. C. Hull, Trans.). In H. Read, M. Fordham, G. Adler, & W. McGuire (Eds.), *Collected works of C. G. Jung:* Vol. 8, *The structure and dynamics of the psyche* (2nd ed., 1969). Princeton University Press. (Originally published 1952)

Mansfield, V. (1995). *Synchronicity, science, and soul-making.* Chicago: Open Court.

Petrarch (1336/1898). The ascent of Mount Ventoux: To Dionisio da Borgo San Sepolcro, in J. H. Robinson (Ed. & Trans.), *Petrarch: The first modern scholar and man of letters.* New York: G. P. Putnam. (Original letter, 1336)

von Franz, M.-L. (1990/1995). Passions of the soul. Ikon Television (Netherlands). Quoted in Mansfield, *Synchronicity, science, and soul-making* (p. 22). Originally broadcast, 1990.

Background and Overview

This volume forms part of the AAPS book series *Advances in Postmaterialist Sciences,* created with the intent to educate scientists, students, and science-minded readers about postmaterialist consciousness research and its implications and applications. We are delighted that the volume is also being co-sponsored by the Scientific and Medical Network's Galileo Commission, with its mission to expand the scope of science beyond the limitations of the currently predominant materialist world view. Our intent is that each volume of the AAPS book series combines rigor and creativity, expresses first person (inner experiences) as well as third person (external observations), and facilitates the evolution of humanity and improved conditions on the planet. The first volume, *Is Consciousness Primary?,* shared both the contributing authors' first-person experiences that transformed their understanding of the primacy of consciousness and the research that supports that perspective. The second volume, *Expanding Science,* offered perspectives on what the unfolding of the postmaterialist paradigm might be like. The third volume, *Spiritual Awakenings in Scientists and Academics,* invited individuals to share the experience of their own spiritual journey, including their spiritually transforming experience(s), the subsequent transformation in world view, values and beliefs and outer changes in their day-to-day life. And the fourth volume was created by the students of AAPS, and included their own essays, poetry and art on the nature of consciousness. Its title is *Consciousness Unexplained.*

The current volume, on synchronicity, grew out of a Galileo Commission working group on the topic that began about a year ago, and includes most of the authors of the contributed essays. Carl Jung invented the word synchronicity, and the term comes from the Greek *syn* (with or together) and *chronos* (time). Thus, synchronicity means together-in-time (Beitman, 2011). One member of our working group, the cultural historian, Richard Tarnas, has written extensively on the topic of synchronicity in his book, *Cosmos and Psyche* (2006), and introduces them as "observed coincidences in which two or more independent events, having no apparent causal connection, nevertheless seem to form a meaningful pattern." He notes that, on occasion, the patterns may strike an individual as being so unusual that it is challenging to believe that the coincidence was a result of chance alone. This is because the events give the clear impression of being carefully and invisibly orchestrated (Tarnas, 2006.)

Since Jung coined the word synchronicity, he and others have adopted a multidisciplinary approach to its study. These areas of research have included physics, biology, psychology, philosophy and spirituality and research on anomalous phenomena (Main, 2004). One of the fascinating areas of research has included the way in which Jung's work, including Main's exploration of synchronicity, responds to a pervasive "dis-ease" in modern culture, the feeling of disenchantment and malaise brought about by our current societal and scientific embrace of materialism as a worldview. Main has proposed that in both human and societal development and evolution, our evolution has moved from a naive enchantment with the world in early cultures and human development, through disenchantment in the age of Enlightenment, as we have focused on intellectualization and rationalization in understanding ourselves and the world around us, through an additional stage of re-enchantment, in which individuals search for an understanding of the deeper mystery, meaning and spiritual reality of life and our relationship to the world around us (Main, 2022). We propose that there is a cultural lag in this developmental trajectory in both the scientific and general academic community, due to the grip of the materialistic worldview in these areas, and how its mechanistic and reductionistic perspective affects the willingness of this community to even entertain the idea of meaningful synchronicity.

In his essay *Science and Humanism* (Schroedinger, 1951), the physicist Erwin Schroedinger, writes that the development of physics has been unsurpassed in transforming our Western outlook on what has often been called the Human Situation. At the time, he forecast that it would take another 50 years before the educated section of the general public would become aware of this change. He made the important observation that the process of cultural assimilation is not automatic: *'We have to labour for it.'* (italics in the original) The same applies to synchronicity, even though such experiences are widely recognized in the general public through their own personal encounters. As we also saw in our volume Spiritual Awakenings, pressure in scientific and academic communities acts as an inhibiting factor.

This current volume, on synchronicity, has been gestating in our collective thoughts for a number of years. David Lorimer notes in his chapter (#13), that in the 1990s, we in the Scientific and Medical Network held a number of working group meetings on what they called 'participatory reality,' inspired by, among others, Richard Tarnas' *The Passion of the Western Mind and The Participatory Mind* by Henryk Skolimowski. Tarnas explains his view in the Epilogue (Tarnas, 1991, pp. 434 ff.) whereby 'the essential reality of nature is not separate, self-contained, and complete in itself, so that the human mind can examine it "objectively" and register it from without. Rather, nature's unfolding truth emerges only with the active participation of the human mind.' It is now evident to us that Descartes was wrong when he split off mind from matter, inner from outer: they are in fact entangled, as synchronicity demonstrates.

In his book *Super Synchronicities,* one of our editors, Gary Schwartz, introduces the concept of "self-science," referring to creatively applying the methods of science to the real-life laboratories of personal life. This methodology has been applied to the creation and thematic content of this volume. The volume consists of essays by scientists and academics describing their own experiences of synchronicity and how these experiences transformed both their worldview and the way they lived their lives. The essays consider possible interpretations/explanations and implications of these synchronistic events for a deeper understanding of consciousness and the interconnected nature of reality in terms of epistemology and ontology. A major theme of the volume is, "What kind of universe makes synchronicity possible?" We anticipate you will be intrigued, and also perhaps feel a sense of participatory resonance and joy, at the answers you find in the essays. We truly believe that this is a fundamentally benevolent, creative and playful universe in which we, as individual expressions of the one Universal Mind, co-create our reality. As you will appreciate, the implications of living your life as a participatory (and playful) co-creator are far-reaching and profound. Life not only happens to us, but we can happen to life. And, as Earl Nightingale tellingly observed, 'Our attitude to life determines life's attitude to us.'

Marjorie Woollacott, PhD
David Lorimer, MA
Gary E. Schwartz, PhD

References

Main, R. (2022). *Breaking the Spell of Disenchantment: Mystery, Meaning and Metaphysics in the Work of C.G. Jung.* Zurich Lecture Series in Analytic Psychology, Vol. 8, Chiron Publications.

Beitman, B. (2022). *Meaningful Coincidences: How and why Synchronicity and Serendipity Happen.* Park Street Press.

Tarnas, R. (2006). *Cosmos and Psyche: Intimations of a New World View.* Viking.

Schwartz, GE. (2017) *Super Synchronicity: Where Science and Spirit Meet.* Param Media.

Chapter One

A Mantis Over My Heart: Experiencing, Interpreting, and Explaining Synchronicity

RODERICK MAIN, PHD

Roderick Main, PhD, works at the University of Essex, where he is Professor in the Department of Psychosocial and Psychoanalytic Studies and Director of the Centre for Myth Studies. His books include *The Rupture of Time: Synchronicity and Jung's Critique of Modern Western Culture* (Brunner-Routledge, 2004), *Revelations of Chance: Synchronicity as Spiritual Experience* (SUNY, 2007), and most recently *Breaking the Spell of Disenchantment: Mystery, Meaning, and Metaphysics in the Work of C. G. Jung* (Chiron Publications, 2022).

Experiencing synchronicity

On the front cover of my book Revelations of Chance: Synchronicity as Spiritual Experience (2007) is an image of a praying mantis, its vivid green profile centered against a white background, its forelegs in their characteristic "praying" posture. The image was chosen by the publisher, State University of New York Press, and they must have liked it, as they also used it for the cover of that year's spring catalogue. What prompted their choice, which otherwise I did not suggest to them, was that my book opens with the following example of a synchronistic experience:

> A professor of biology, Adolf Portmann, was delivering a lecture that he intended to conclude with a story about a praying mantis. Just as he was about to broach this subject, a praying mantis flew into the lecture hall through an open window, circled around Portmann's head, and landed near the lectern lamp, to the effect that the insect's [forelegs] cast on the white wall behind him a huge shadow in the form of the arms of a praying man. (Main, 2007, p. 1, based on Quispel, 1995, p. 17)

Powerful though it is, I did not further discuss this example in my book (or in the PhD on which the book was based), since I had become aware of it late in the day and had opened with it mainly just because of its similarity, in both form and content, to C. G. Jung's (1952/1969b) famous example of the scarab beetle (paras. 843, 845): In both cases, an unusual insect dramatically enters the room through a window simultaneously with the primary experiencer's mentioning (or being about to mention) precisely that insect.

I knew from previous reading that the praying mantis is a richly symbolized and mythologized creature, and I was aware that, besides Portmann, at least two other notable writers with connections to Jung—Laurens van der Post (1975) and Joseph Campbell (see Grof, 2006, pp. 8–10)—had experienced synchronicities involving the praying mantis. But I had not explored this symbolism or these experiences in any depth, as I did not, up to the time of publishing my book, feel a particularly strong personal connection to the symbol. However, after the publisher's foregrounding of the praying mantis on the cover of my book and in view of these other associations and the primary meaning of the ancient Greek word "mantis" as diviner, seer, prophet, I did begin to wonder whether this creature, like the scarab (see Main, 2013), might be specially related to synchronicity or even be an apt symbol of synchronicity itself.

Then, in late September 2009, I travelled to Shiga Prefecture in Japan to join my wife and our two young (pre-school) children for a visit to my wife's parents, as well as, opportunistically, to give some talks to members of the Association of Jungian Analysts of Japan and to students at the University of Kyoto. I had brought with me a copy of *Revelations of Chance* as a gift to my parents-in-law—despite that neither of them can fluently read English!

One day during our stay, my wife and I, our children, and my wife's mother visited the nearby shrine at the foot of Mount Mikami, a prominent geographical feature of the area rising 432 meters above the rice plains. The shrine compound comprises several buildings, and I was apart at one of these when my wife called me over excitedly. She and the others were standing in a circle looking at something on the ground. When I joined them, I saw that it was a praying mantis, the first living one I had ever seen, and I was impressed by both its unmistakable appearance and its size (perhaps four inches long).

As we watched, the mantis stepped, very slowly and precisely, towards where I had joined the circle. It continued right up to my shoe; it even began to climb onto my shoe. Then, all of a sudden, unbelievably quickly, it scurried up my leg and torso and came to rest on the left side of my chest—over my heart, as I realized afterwards.

Somewhat apprehensively, I peered down at the mantis, which, now motionless, seemed in turn to be gazing at me. We eyed each other for a few moments. Then, since it showed no sign of moving off by its own decision, I tried to remove it by blowing, first gently, then more vigorously. As it still didn't move, my mother-in-law, seeing my intention, lightly flicked it away.

Interpreting synchronicity

This experience matches Jung's definition of synchronicity as involving a psychic state and a physical event connected acausally through meaning (1952/1969b, paras. 849–850). The *psychic state* was my growing interest in the symbol of the praying mantis after my publisher's use of its image. The physical event was the appearance and behavior of the actual praying mantis at the shrine. The event was acausal in that there was no plausible way by which, in terms of efficient causes, either my interest in the symbol of the mantis could have caused the appearance and behavior of the actual mantis or the appearance and behavior of the actual mantis could have caused my prior interest. Finally, the psychic state and physical event were connected through meaning, on several levels: (1) they expressed parallel content in notable detail; and their coincidence (2) was accompanied by a powerful emotional charge or sense of numinosity, (3) had a personally transformative effect on me, and (4) evoked a complex field of significance that transcended, even if it may partly have included, my personal concerns (see Aziz, 1990, pp. 64–66, 75–84; Main, 2022, pp. 78–81). I was partly aware of each of these levels of meaning at the time of the experience, though some aspects, especially related to the last two, emerged more fully only later once I was able to reflect on the experience.

Paying attention to the parallel details that establish the synchronicity can help to provide a basis for exploring its possible meaning more deeply. For example, in my experience, the content expressed by the psychic state and physical event was not just the image of a praying mantis. There were also several more specific associations: The praying mantis was associated with the *sacred* in that my psychic state of interest related to the cover of a book on *spiritual experience,* while the physical event occurred at a *shrine;* and it was associated with *synchronicity* in that my interest in the mantis was due to its having been the focus of *impressive synchronistic experiences* about which I had read, while the appearance of the mantis at the shrine precisely was a *synchronistic experience.* Since these specific details were salient, it would seem reasonable for my interpretations to focus on those aspects of the symbolism of the mantis that may relate to the sacred and to synchronicity.

The emotional charge or sense of numinosity of the synchronicity constitutes a nonrational or at least pre-discursive form of meaning. At one level, this could include the "ontological shock" (Kripal, 2022, p. 15) that synchronistic experiences can deliver, the sense that something has happened that is incomprehensible in terms of our ordinary understanding of reality. At another level, the sense of numinosity, of something at once weird and profoundly meaningful, can both spur discursive inquiry into the meaning of synchronicities and guide our search for specific interpretations, enabling us to sift those interpretations that feel at least partially adequate from those that feel less so.

At the level of personal transformation, I found that my experience involving a symbol which was already deeply implicated with synchronicity served to reground me in the topic during a period when, because of other academic duties, I was finding less time for research on synchronicity. More specifically, at a time when I was under tacit external pressure, because of the secular character of my employing institution, not to pursue the spiritual implications of synchronicity, this synchronistic experience fortified me in the way it revolved around a creature whose very name, in English as in many other languages, is charged with spiritual associations, not to mention the religious myths about it in many cultures and the fact that it appeared to me at a shrine. Not least, because of the meaning of "mantis" as "diviner," the incident has reinforced my sense of the importance of divination as a field for exploring and applying synchronicity. That not just incidental but central importance should be given to these associations with synchronicity, the sacred, and divination is further suggested by the mantis's having appeared centrally on the book cover and then at the center of a circle of onlookers and even more prominently on my chest. Its resting over my heart and exchanging a momentary gaze makes me consider how predominantly detached and intellectual my inquiries into synchronicity and the sacred have often been and prompts me to attend more to affective, participatory, and even unitive approaches to understanding—to learn to see "with the eyes of the heart," as it is expressed in the *Hermetica* (Copenhaver, 1992, pp. 17, 24). Finally, from a different perspective, perhaps the way the synchronicity unfolded, where it was not I but my wife, children, and mother-in-law who noticed the mantis and whose circle around it I joined, was a hint that I should aim to integrate better my work and research life with my family life and to value the latter more.

At the level of transpersonal significance, the praying mantis is a creature to which has accrued a rich field of mythological and folkloric meanings across many cultures, from the most ancient to the most modern. Most famously, perhaps, the mantis has been celebrated by Laurens van der Post as one of the most important deities of the first peoples of southern Africa (1975; see also Bleek et al., 1924). But it has also long been known to have had religious connotations and been associated with supernatural powers among, for instance, the ancient Nubians, ancient Greeks, Turks, Arabs, and inhabitants of Provence (*Encyclopedia Britannica,* 1911, p. 606). However, not all the associations are immediately spiritual. In Japan, where my experience took place, the praying mantis is named *kamakiri,* literally "sickle cut," or "cuts like a scythe," a truer rendering of the deadly function of those ostensibly praying forearms. In China, various forms of martial art were developed based on observing the behavior of the mantis (Kripal, 2024, citing Farrer, 2013). In modern Europe, the praying mantis was an important symbol for surrealist writers and painters, such as Paul Éluard, Salvador Dali, André Masson, and Max Ernst, who seem to have been particularly fascinated by the female mantis's practice of cannibalizing her male partner after intercourse (Kripal, 2024, citing Pressly, 1973). And, at a level of weirdness that makes my experience seem positively ordinary, multiple accounts are emerging in contemporary alien abduction literature, particularly in the United States, of experiencers claiming to have encountered giant mantis-like entities (Kripal, 2024). These associations testify to the fascination that the mantis continues to exercise over the human psyche. At present, I am not sure what to do with some of the associations, other than simply to expand the field of possible transpersonal significance of my experience to include them. Overall, however, the associations of the mantis with synchronicity and the sacred seem unmistakably pronounced.

It may be worth mentioning, too, that some of the above information about the mantis only came to my attention, unsought, after I had begun writing the present account of my experience. I had taken the half-written account with me to work on while participating in a conference on "Edges

When one recognizes and engages hermeneutically with an archetype that has irrupted into one's life through a synchronistic experience (or in some other way), one can get drawn, to however small a degree, into the process of individuation and thus towards greater, even if in practice never complete, realizations of wholeness.

Despite Jung's coyness about engaging in metaphysics, it is clear that these holistic formulations have ontological implications. They affirm the reality of a transempirical, transcendent, or sacred order of reality from which the archetypal meaning of synchronistic experiences stems, meaning which Jung also variously calls "objective," "a priori," and "self-subsistent" (1952/1969b, paras. 915, 942, 944). Jung's holistic formulations also implicitly affirm an interplay or paradoxical unity of this transcendent order with its immanent expressions in the psychic and physical contexts of empirical reality.

This all suggests to me that Jung's understanding of synchronicity can best be framed within a spiritual metaphysical framework—specifically, I would argue, within the framework of panentheism. As I understand it, panentheism affirms the reality of the divine and of a relationship between the divine and the world which balances transcendence and immanence through envisaging the divine as not separate from the world, as affected by the world, and as more than the world (Main, 2022, pp. 109–119). Historically, this metaphysics has been expressed by, for example, Hermetic and Kabbalistic worldviews, as well as implicitly, I have argued, by Jungian psychology, whose panentheistic structure consists in Jung's effectively equating the unconscious with the divine and consciousness with the world and then viewing the unconscious as not separate from, as affected by, and as more than consciousness (pp. 112–119). As an overarching framework, panentheism can also host, it seems to me, various other, more philosophical and scientific frameworks that have been deployed to make sense of synchronicity, such as emergent panpsychism (Cambray, 2009), Whiteheadian process philosophy (Haule, 2011), objective idealism (Kastrup, 2021), and dual-aspect monism (Atmanspacher & Rickles, 2022)—though this intuition still needs to be worked through in detail. Meanwhile, from a panentheistic perspective on synchronicity, we can suppose that the transcendent meaning expressed by the archetype as such is not separate from, is affected by, and is more than the immanent meaning expressed by the archetypal images manifesting in the psychic states and physical events of a synchronistic experience. More concisely put, in synchronicity the divine can manifest within our empirical experience.

It was perhaps inevitable that my theoretical reflections would highlight the relationship between synchronicity and the sacred, considering that my interpretations of the content of my synchronistic experience with the mantis pointed in this same direction. Be that as it may, I should like to end by noting that all these epistemological and metaphysical reflections, as well as all the interpretations in terms of personal transformation and transpersonal significance mentioned earlier, are basically an imperfect representational unfolding of what was—and still is—implied in the momentary synchronistic experience itself, from which I started.

Conclusion

At Mikami shrine
A mantis over my heart—
Its eyes meeting mine.

References

Atmanspacher, H. & Rickles, D. (2022). *Dual-Aspect Monism and the Deep Structure of Meaning*. Routledge.

Aziz, R. (1990). *C. G. Jung's Psychology of Religion and Synchronicity*. State University of New York Press.

Bleek, W., Lloyd, L., & Bleek, D. (1924). *The Mantis and His Friends: Bushman Folklore*. T. Maskew Miller/Blackwell.

Cambray, J. (2009). *Synchronicity: Nature and Psyche in an Interconnected Universe*. Texas A & M University Press.

Copenhaver, B. (Ed.). (1992). *Hermetica: The Greek* Corpus Hermeticum *and the Latin* Asclepius *in a new English translation with notes and introduction*. Cambridge University Press.

Encyclopedia Britannica (1911). Mantis (Vol. 17, p. 606).

Farrer, D. (2013). Becoming-animal in Chinese martial arts. In P. Dransart (Ed.), *Living Beings: Perspectives on Interspecies Engagements* (pp. 145–165). Bloomsbury.

Grof, S. (2006). *When the Impossible Happens: Adventures in Non-Ordinary Realities.* Sounds True, Inc.

Haule, J. (2011). *Jung in the 21st Century, Volume 2: Synchronicity and Science.* Routledge.

Heath, S. (2012). Honoring the mantis between our toes. *Psychological Perspectives 55:* 331–341.

Jung, C. G. (1969a). On the nature of the psyche. In H. Read et al. (Eds.), *The Collected Works of C. G. Jung: Vol. 8. The Structure and Dynamics of the Psyche* (2nd ed., pp. 159–234). Routledge & Kegan Paul. (Original work published 1947, revised 1954).

Jung, C. G. (1969b). Synchronicity: An acausal connecting principle. In H. Read et al. (Eds.), *The Collected Works of C. G. Jung: Vol. 8. The Structure and Dynamics of the Psyche* (2nd ed., pp. 417–519). Routledge & Kegan Paul. (Original work published 1952).

Jung, C. G. (1970a). *The Collected Works of C. G. Jung: Vol. 14. Mysterium Coniunctionis: An Inquiry into the Separation and Synthesis of Psychic Opposites in Alchemy* (2nd ed.; H. Read et al., Eds.). Routledge & Kegan Paul. (Original work published 1955–1956).

Jung, C. G. (1970b). The undiscovered self (present and future). In H. Read et al. (Eds.), *The Collected Works of C. G. Jung: Vol. 10. Civilization in Transition* (2nd ed., pp. 245–305). Routledge & Kegan Paul. (Original work published 1957).

Jung, C. G. (1977). Religion and psychology: A reply to Martin Buber. In H. Read et al. (Eds.), *The Collected Works of C. G. Jung: Vol. 10. Civilization in Transition* (2nd ed., pp. 663–670). Routledge & Kegan Paul. (Original work published 1952).

Kastrup, B. (2019). *Decoding Jung's Metaphysics: The Archetypal Semantics of an Experiential Universe.* Iff Books.

Kripal, J. (2022). *The Superhumanities: Historical Precedents, Moral Objections, New Realities.* The University of Chicago Press.

Kripal, J. (2024). *How to Think Impossibly: About Souls, UFOs, Time, Belief, and Everything Else.* The University of Chicago Press.

Main, R. (2004). *The Rupture of Time: Synchronicity and Jung's Critique of Modern Western Culture.* Brunner-Routledge.

Main, R. (2007). *Revelations of Chance: Synchronicity as Spiritual Experience.* State University of New York Press.

Main, R. (2013). Myth, synchronicity, and re-enchantment. In L. Burnett, S. Bahun, & R. Main (Eds.), *Myth, Literature, and the Unconscious* (pp. 129–146). Karnac.

Main, R. (2022). *Breaking the Spell of Disenchantment: Mystery, Meaning, and Metaphysics in the Work of C. G. Jung.* Chiron Publications.

Main, R. (Ed.). (1997). *Jung on Synchronicity and the Paranormal.* Routledge.

Pressly, W. (1973). The praying mantis in surrealist art. *The Art Bulletin,* 55(4): 600–615.

Quispel, G. (1995). Gnosis and psychology. In R. Segal et al. (Eds.), *The Allure of Gnosticism: The Gnostic Experience in Jungian Psychology and Contemporary Culture* (pp. 10–25). Open Court.

van der Post, L. (1975). *A Mantis Carol.* The Hogarth Press.

Chapter Two

Synchronicities: Staying True to Your Soul's Blueprint

JOAN WALTON, PHD

Joan Walton, PhD, received a first degree in Social Theory and Institutions (University of Bangor, North Wales) and MSocSci in Race and Education. Following University, she travelled extensively in India and South East Asia, worked for several years in social work, then founded and ran an independent centre for action research and education. Following completion of her PhD in 2008, she had a career change, and entered the academic world, initially at Liverpool Hope University, and now part-time at York St John University in the School of Education. Her main research interest is investigating nonmaterialist ideas about consciousness, and exploring what relevance these ideas have for education. With colleagues, she has developed an international network of academics and educators, interested in evolving Consciousness Education across all age ranges, in diverse educational settings.

Introduction

I have experienced synchronicities throughout my life, both minor, in the sense that they could easily be put down to 'chance', (e.g. the phone rings just after you are thinking of the person who is phoning); and major, which appear to be far beyond that which could be counted as coincidental.

I start this chapter by giving an account of three synchronicities. The first one I have told on many occasions, mainly because it created an exceptional transformative shift in my own consciousness, confirming (for me) that we live in a universe much more mysterious and meaningful than our limited embodied experience might have us believe. The second happened to my daughter, suggesting an ongoing connection with loved ones who have passed on. The third is more recent, but is an example of a particular kind of synchronistic event which affirms the sense of belonging to a reality that is truly wonderful (a word that is significant in the synchronicity itself), and in 'mundane' life, might be viewed as magical. It feels to me that the unlikeliness of each of these events radically challenges any claim that they could have transpired by chance.

First synchronistic experience

When my daughter, Rachel, was about six years of age, we arranged to go on holiday to France with friends who had two young children. We had a four-berth caravan, and a tent that slept six. The friends were to borrow our tent. As we lived in the Midlands, and they were travelling from Dorset, we arranged to meet them at the port, prior to taking the boat over the channel.

We drove down overnight, and stopped in a lay-by for an early morning breakfast. The tent was in the caravan, and as was our custom, we took it out, and put it under the caravan to give us room inside. Whilst eating, a minibus pulled up behind us. I realised that my daughter was paying them a lot of attention out of the window, and saw that one of them was relieving himself close to our van. I closed the curtains, telling Rachel to get on with her breakfast. When we had finished, we went to fetch the tent—to find it had gone. The young men had apparently not only relieved themselves, but had relieved us of our belongings! This put us in a fairly dreadful situation. Our friends were dependent on us for their accommodation. We were none of us well off in those days, and could not afford to either buy another tent, nor pay for alternative accommodation. The caravan was hardly big enough for us, far less sleep four additional people.

At our wits end, we decided we might as well go to the campsite we were booked onto, as there seemed no better alternative. When we reached there, I asked the person at reception whether she knew of anywhere we could hire a tent very cheaply, as ours had just been stolen. The woman gave us a strange look, and said we were in luck. A family had left the day before, and had given her their tent prior to leaving. Apparently they were to buy a new tent for their next holiday, and would have no use for their old one. They thought she might be able to find someone who would have use for it. The woman had not had this happen to her previously, and wasn't quite sure what to do with it—so when we turned up, she was glad to have it taken. That was coincidental enough—but the real shock came when we saw the tent. Ours was about twenty years old, blue and orange with two sleeping compartments, good quality but well worn with two or three tears in different places. This tent was virtually identical—same age, colour, size, condition—to all intents and purposes, the same tent. It wasn't—the tears were in different places. But if you didn't know the other tent as we did, and put the two together, you couldn't have told the difference. We were stunned.

Throughout the time from the theft of the tent, to reaching the destination, we had gone through numerous possibilities about what we could do, but had come up with no answers. We had no spare money. I was feeling incredibly anxious, and guilty about letting our friends down. It seemed our holiday was heading for disaster. Then, at a peak of worry and lack of possible solutions, this seemingly miraculous event occurred, dissolving in a moment all the concerns, and leaving us once again with the prospect of a happy holiday ahead.

Second synchronistic experience

My partner Jerry and I had just bought a new home. My daughter Rachel was so excited and had a calendar in which she marked off each day till we would begin our new life together. Our new life together was to begin on a Friday, and the Thursday night just prior, I learned that Jerry had a heart attack and died. I woke my daughter up at 2 am to tell her. I realized that we were indeed about to begin a "new life" but now because of the death of Jerry it was not going to be the new life we had imagined.

Prior to Jerry's death I had been convinced (reinforced by the synchronistic experience of the tent) that there was a power greater than myself that was guiding me. I believed that this power was a benevolent purposeful force at work in my life, and in the universe. But now I questioned this. How could the death of Jerry be consistent with this belief?

What possible meaning could there be in his death? Was his death in fact evidence of the opposite belief, that there is no benevolent power? If something this awful can happen, then perhaps it proves that there is no benevolent power giving meaning in life.

But something kept nagging at me, and so I spent considerable time contemplating this dilemma. I began to consider that perhaps there was some meaning to this awful event, but it was beyond my ability to understand, at least right now. I began to think "I need to remain open to the possibility that there is meaning in this event." I created the hypothesis that "Jerry's death was meaningful, that there was a purpose behind it, and one day I would understand it. I would act as if this were true." As time went on, I felt that the consciousness that was Jerry was around in some mysterious way, supporting both Rachel and me.

Many years after the death of Jerry, my daughter, Rachel, now 17, was driving her car over a bridge out in the country, with her music playing loudly as normal. Because it was a single lane bridge it was important to slow down and make sure that no cars were coming in the opposite direction. Rachel slowed down and then tried to accelerate over the bridge but her foot wouldn't move on the accelerator. At that moment a car came speeding over the bridge, not slowing down or looking for other cars and zoomed past her. Because of her inability to move her foot and accelerate her car over the bridge, Rachel had avoided a head on collision.

Rachel was too preoccupied with the shock of what had happened, and the potential disaster that had been so narrowly averted, that she did not think anything particular about this for a few minutes. But as she drove on, she realised the significance of what had just happened. Her life had been saved, because her foot suddenly froze, and she couldn't press the accelerator. A strong sense of Jerry pervaded her consciousness at the same time. What possible explanation could there be for the concurrence of these two aspects of her experience, when she was so occupied with what was happening in the present moment, and no time to think and reflect?

It feels to me to be a strong affirmation that there is indeed a wider reality, which we are not able to see directly from a limited human perspective, but is indeed present with us, and is supporting us on this life journey, if we are open to that possibility.

Third synchronistic experience

Just before Christmas 2022, I was in my house, waiting for my two young grandchildren to arrive. They were coming to watch 'The Snowman', a film based on the book of the same name, written by Raymond Briggs. It was 3.50 pm, and I put the TV on, ready for The Snowman to start at 4 pm. Just at that moment, a text message arrived from a friend in America. I had previously sent him a photo of the two girls preparing for Christmas. He wrote a message which said that the picture I sent reminded him of a scene near the end of the old black and white film 'It's a Wonderful Life'.

And—staggeringly—what was facing me on my TV at that moment, was the end of 'It's a Wonderful Life'.

I looked at the screen, and looked again at the text. It seemed beyond belief. I then took a screenshot of the TV guide, showing that *It's a Wonderful Life* did indeed immediately precede *The Snowman.* I'm not sure it's possible to work out the statistics of that happening through chance. The question is, is there a rational explanation for it? The frequency of synchronicities suggest we need to accept a way of understanding reality that lies beyond rationality.

Personal significance of synchronicities

I have always accepted synchronicities as indicators of a reality that exists beyond that which we experience with the five senses; or to use William James' framing, there is a "MORE which is operative in the universe ... and which (you) can keep in working touch with". But there were also signs that life could be cruel, evidenced by the amount of suffering that exists. What kind of explanation could account for both the magical and the misery? I lived with an ambivalence between certainty that the world had to be meaningful, and the possibility that some kind of great cosmic tragedy was being played out that was tricking us all.

However, the synchronicity with the tent changed that. It overwhelmed me. I had to go off on my own for a while, just to process its significance. My main and sustaining feeling was, that this was a message to tell me that, if ever I had doubted there being a benevolent Power that lay behind our daily experience, I should not do so again. I absolutely knew that to put this down to a meaningless coincidence would be to trivialise something wonderful. I do not use the word 'sin' often, and in fact (reacting against my strong Christian upbringing), I do not like it as a word; but I found myself feeling at a deep level that it would indeed be 'sinful' to dismiss this as chance. The intuitive arising of this term was not aimed at making me feel guilty; but more to indicate that it would be a travesty of life's meaning to minimise the worth of this experience, or to doubt the message it was sending me.

Theoretical explanation

In my theoretical explanation of events of this nature, I have tended to summarise a version of Carl Jung and Wolfgang Pauli's interpretation, by saying that, if you are open to possibilities, and accept the unity of all existence, there will be a coincidence of psychic and physical reality, which will present as synchronicities. In my PhD thesis, when exploring this, I wrote as follows:

"Carl Jung asked how, in a world supposedly governed by the law of causality, could there be phenomena which so clearly violated that law? His answer was to doubt, not the veracity of the phenomena, but the universal validity of the law. Causality is only one principle. If the law of causality couldn't accommodate the existence of certain facts associated with the workings of the mind, among them telepathy and precognition, then that law must rest on a distorted or, at best, partial view of reality, and was itself in need of some rethinking.

Jung was encouraged to pursue this stance by new developments in 20th century physics. He knew Einstein's relativity theory had challenged all old notions of space and time which were part of the causal framework, and that the unpredictable events described by quantum physics seemed more relevant to understanding the mechanics of the psyche. If the universe were to be understood fully, this task would be achieved by both physics and psychology transcending their own disciplinary limitations and moving forward together with shared insights.

He developed an alliance with Wolfgang Pauli, the Nobel prize-winning quantum physicist, who also believed there was a reality lying beyond that which could be explained by cause and effect. Parapsychology was seen to act as a natural bridge between physics and psychology; and Pauli hoped that in working with Jung he might find a way to express on the larger scale everyday level of reality some natural extension of the very small scale, quantum mechanical phenomena he had helped to discover.

Jung and Pauli supported the view that there is an absolute spaceless, timeless cosmos in which both the psyche and the material universe are manifest. Pauli argued that this cosmos has an order of its own, independent of human will, human perceptual categories or our supposed laws of causality. Within this viewpoint, all accepted boundaries between the knower and the known break down, and mind and matter come to be seen as extensions of each other.

Jung called this somewhat mystical absolute 'transpsychic reality', and argued that, in a realm beyond our conscious psyche with its divisions between mind and matter and its causal perceptions manifested in space and time, there is a timeless unity in which past, present and future merge, and where matter and the psyche are but alternative manifestations of a single reality."

Ontological understanding of reality

As the years have progressed, I have been less concerned about discovering a precise academic explanation, and more interested in exploring the implications for how I perceive and live my life. In other words, what is the potential, both for ourselves as individuals, and for humanity as a whole, if we live each moment in the knowledge that we are integral to an expanded reality that is powerful beyond our conception? And what are the implications for our ontological understanding of reality?

I am very aware that any conception I have of the universe may change as a consequence of further experience. So I have a lightly held 'living theory', open to modification. As things stand, though, my ontological view is that all living beings are unique expressions of a Universal Consciousness, which is Infinite and Eternal, with the core qualities of Love and Creativity, and which has Intelligence.

This Infinite and Eternal reality contains all that is and ever has been: there are no boundaries, no separation. Nothing is ever lost. If we are open to what is held in that wider reality, we are able to access the learning and wisdom about living as embodied beings that has been gathered through the ages.

The ultimate purpose of this Universal Consciousness is to express itself as purely as possible in a physical form.

In aiming to find a theoretical framing that makes sense of my views, I find Bernardo Kastrup's ideas of analytic idealism resonate strongly.

How does your ontology influence your idea of meaning?

Within my ontological framing, the experience of meaning is what is felt when one is in tune with the qualities and purpose of the Universal Consciousness.

So if the ultimate purpose of human existence is to be a pure expression of the Universal Consciousness (that is, to creatively express pure Love in the world, in whatever way is right for the individual involved—which could be in relationship with others, or through music, art, or any other form of loving, creative action), then a sense of meaning will be felt. And pursuing a deep and authentic sense of meaning is a guide to telling us how to contribute to an evolution of consciousness, to the story of the Universal Consciousness realising itself in embodied form.

What are the implications of this ontology for my understanding of synchronicity?

It seems to me that synchronicity occurs when one is living in a way that is in tune with the purpose of the universe in a way that feels meaningful; or alternatively is seeking at some deep level to find a way to do so. Each person is possibly born with a unique purpose, and synchronous experiences may be indicators that one's actions are in line with fulfilling that purpose. In other words, synchronicities are a form of communication between embodied living and wider dimensions, conveying a message that life is being lived in resonance with individual and universal purpose.

The message I have learned can be summarised as follows:

- Stay open to possibilities.
- If you are open to miracles, miracles will happen.
- Don't take any 'coincidences' for granted! Take notice, and be responsive.

Conclusion

Personally, although I think it is intriguing to research synchronicity from a third person perspective, I am reminded of the double slit experiment in quantum physics, where merely the presence of the observer changes the behaviour of the photon. If I start studying synchronicity 'objectively', and seek to understand the conditions under which it occurs, will that not actually alter both the nature and frequency of the synchronicities? This is a question I consider, rather than a fixed viewpoint I hold. But it leads me to think that, if we wish to experience and comprehend more profound synchronicities, it benefits to pay attention to staying true to your soul's blueprint, and following where that takes you. If there is a direct relationship between the authenticity of how we live our lives, and the quality of the synchronicities that occur, then perhaps more can be learned about the nature of that relationship through the 'thick' descriptions that emerge when first person narratives are shared. In which case, this book will make a valuable contribution to our understanding of synchronicity.

References

James, W. (1902) *The Varieties of Religious Experience: A Study in Human Nature,* New York: Longmans, Green.

Jung, C.G. (1960/1973) *Synchronicity: An Acausal Connecting Principle,* USA: Princeton University Press.

Kastrup, B. 2019. *The Idea of the World, A Multi-Disciplinary Argument for the Mental Nature of Reality.* Winchester: Iff Books.

Chapter Three

Synchronicity Unveiled: Bohm's Holomovement and Indra's Net

WILLIAM KEEPIN, PHD

William Keepin, PhD, is a mathematical physicist with extensive training in contemplative meditation and transpersonal psychology. His research on sustainable energy to mitigate climate change influenced international environmental policy, and he was a whistleblower in nuclear science policy. Will co-founded Gender Equity and Reconciliation International which has conducted 300 trainings in 14 countries on gender equality, and also Satyana Institute which organizes conferences to bring contemplative mystics together across the religions and science. His books include *Belonging to God: Science, Spirituality, and a Universal Path of Divine Love* (2016), and *Gender Equity and Reconciliation: Thirty Years of Healing the Most Ancient Wound in the Human Family* (2023).

Experiences of synchronicity

As a graduate student in quantum physics in the late 1970s, I was thoroughly steeped in the scientific mindset. I felt fortunate to have a research assistantship with one of my senior professors. We were researching the quantum mechanics of air, by generating numerical solutions of the Schrödinger equation to calculate the vibrational energy levels of diatomic molecules.

One fine weekend I took a day hike with two friends at the Valles Caldera, a dormant volcano in the mountains west of Los Alamos, New Mexico, where I had grown up. We hiked across the high caldera valley at 11,000 feet, toward the rim on the other side. One of my friends had personal connections with the Native American community, and she brought along some high quality plant medicine. So this particular experience was facilitated by this powerful psychedelic.

We spent an absolutely idyllic day, communing with every aspect of nature we encountered. In our heightened state, I recall vividly perceiving the exquisite uniqueness of each life form. At one point we came upon a vast grove of Aspen trees, and the presence of those trees was awe inspiring, as we felt somehow warmly embraced and welcomed into this family of vibrant Aspen trees. Words are inadequate, but as can happen in these expanded states of awareness, I felt that I was literally experiencing the essence of these living beings merging into my own. I had done psychedelics several times before, so this was familiar experiential territory, but on this day, it was particularly strong, and exceptionally ecstatic and joyous.

We hiked for over three hours to reach the far side of the caldera, and then began hiking up the mountain toward the far rim. At one point unexpectedly, I started to feel a bit nauseous, and a little fearful. I hadn't felt anything like this all day, and I was a bit surprised this was coming on so many hours after taking the plant medicine. So I just noted it, and figured it would soon pass. It didn't. As we continued climbing up this ridge, the feelings gradually intensified, and I began to sense a kind of dread. I became very quiet, staying centered and waiting/hoping this experience would pass. I was a bit concerned that I might descend into a "bad trip," and I was working to stay centered and calm. All three of us had fallen silent for a while, and then one friend finally broke the silence, to announce that she wasn't feeling very well. The other friend immediately said that she felt bad too—and we suddenly realized that all three of us were feeling the same way.

This was surprising, and then we also realized that as we ascended this mountain toward the rim, these feelings were getting worse, the closer we got to the top. We were all three having the same shared experience, and it felt somehow related to whatever was up there on or beyond the rim. We became curious about what was on the other side of that rim. As we approached it, we did so with some trepidation, aware that we might find something disturbing. We stopped to take stock, and discussed possibly turning around, but we all three decided that we needed to continue, and find out what was up there.

We continued hiking up, and the negative heavy feelings continued to grow stronger with every step. Just as we crested the top rim, we came upon a huge clear cut of the forest. Stumps everywhere! The impact was overwhelming and devastating. Each stump was still slightly moist, savagely seared by the saw. The psychological impact felt like each stump had been a human being standing there, who was severed at their ankles. Large sections of trunk and branches were lying around, which looked and felt like bloody body parts strewn across a battlefield.

All three of us burst into tears, and began to wail uncontrollably. The pain of the trees seared into our bodies, we could also feel the ruthless insensitivity of the consciousness with which this had been carried out—raping and ravaging the forest, and hauling the dying trees away. It literally felt like we were the first on the scene of a bloody battlefield, where our beloved friends had just been brutally massacred.

We continued to cry for a while, and then moved into spontaneous apologizing to the trees. Stricken with grief, we supported one another emotionally, while also feeling deeply complicit as part of the offending species. We spent well over an hour there, and when we finally left we were exhausted, but we still had to make the whole hike back. We reached our car at dusk.

This was a life-changing experience for me, and it was part of my 'conversion' into an environmental scientist, and a spiritual seeker. I had earlier had another conversion, knowing that I would never have anything to do with nuclear weapons. Now I knew I must do something about the destruction of the earth. Not long afterward, I began meditation practice in earnest.

I contemplated this experience deeply for some time. In many ways, this single experience jettisoned the entire mechanistic scientific worldview right out of my mind, for good—because I knew it was incomplete, shallow, too mechanistic, and in some essential way, mistaken. In this encounter, all three of us knew something, perceived it directly, yet without any access to any physical data or inputs. We could neither see, smell, taste, hear, nor touch that clear cut—but we *felt* it within our own being, from at least half a mile away. We felt an invisible force field of radiating agony, of the suffering of those trees and the Earth. We had each felt it independently, before we realized that we were all feeling it together.

According to my scientific training, as I understood it then, this cannot happen. You can't perceive things that you don't see, hear, touch, taste, or feel physically. But we *did!* I knew that this was not a signal-receiver type of communication. It was a knowing by mergence of awareness-es. The awareness of the trees in their suffering, merged with my awareness and became one shared knowing, human-and-tree-as-one, and this all happened at a substantial distance from the clear-cut site. Perhaps one could explain this in terms of electric and magnetic force fields or whatever, but it felt like there was another dimension to it that was completely outside of normal scientific transfer of information.

We had spent the entire day communing with the beloved trees and plants and the small wildlife of nature. And then we came across this scene, and it hit us like a horrific slaughter of innocents. Those trees were living beings, egregiously slain with savage blades wielded by our human brethren. There was a strong moral impact, a kind of naked unmasking of the shadow side of our entire western civilization. The ruthless and appalling insensitivity of the agents responsible for the clearcut was almost as painful as the fallen trees themselves. We all knew in our hearts that this same insensitive murderous ravaging of nature is carried out daily, all across the planet, in our relationship with the living Earth. As the indigenous Kogi elders once lamented, "If they knew the Earth could feel, they wouldn't be doing this."

Of course, one could just dismiss this example as 'oversensitivity' induced by psychedelics. But the fact is that we had an objective perception of something real from a long distance away. We effectively perceived telepathically, through a kind of epistemology of the heart, about which science still knows precious little but mystics have known for millennia. I also realized that the machines that had carried out these heinous actions were invented largely by that very science that I had been studying so earnestly all those years in college and grad school.

Another experience of synchronicity

I first met Richard Tarnas in 1989 when he came as Guest Faculty to the three-year intensive training I was doing with psychiatrist Stanislav Grof. Stan had asked Rick to introduce us to astrology, and show us how historical events and deep psychotherapeutic experiences are correlated with the movements of the planets. I was deeply skeptical at first, and insisted upon seeing compelling, objective data. Although I was open minded, and by that time I had had many powerful spiritual and mystical experiences in meditation, the scientific worldview and quantum physics were deeply ingrained in me. I was open to lots of radical ideas, experiments, and possibilities, but ... astrology? It seemed way too far beyond the wildest pale to be taken seriously. It took me considerable time to even embrace the astrological data, and I was asking non-stop questions, and essentially demanding proof at every step.

I was going through the same process that psychiatrist Stanislav Grof had gone through a couple decades earlier. He had been investigating various standard personality tests like Meyers-Briggs and Jungian typology, to better understand the experiences his clients were having in LSD psychotherapy. Stan was finding no useful results whatsoever with these standard personality models, and then an astrologer approached him, and revealed to him striking patterns of correlation between his clients' experiences and their outer planet transits. Truly astounded, and against every scientific bone in his body, Stan began to explore astrology.

I was now undergoing this same process, totally 'taboo' for a scientist. Yet I began to see the undeniable correlations. I also began to witness the remarkable correlations between clients' deep experiences in non-ordinary states of consciousness facilitated by Holotropic Breathwork, and their outer planetary transits. This deepened over the next several years, as I began tracking the transits of clients in our work, and began to see striking correlations between their transformative experiences and their outer planet transits. These experiences began to expand and utterly transform my scientific worldview, into a far deeper paradigm than even quantum physics and relativity theory had gifted me. Over time, astrology unveiled itself as the quintessential science of synchronicity. Grof later identified eight key ways that "decades of systematic research of holotropic states have generated vast amounts of data that undermine basic assumptions of materialistic science and bring supportive evidence for astrology" (Grof, 2009, p. 53).

A key moment of synchronicity for me came early on in this process, when I examined the planetary transits of a close family member who had had a psychotic break several years earlier. This had been a huge shock to the whole family, because she had never exhibited any mental health issues throughout her life, and we had always wondered why and how this had happened, and so suddenly. At that time she had transiting Pluto conjoined her natal Neptune, which is squared to her natal Mars-Uranus conjunction. Ebertin's well-known book *Combinations of Stellar Influences* specifically names mental health challenges for the Mars/Uranus/Neptune triple interaction. So there she was, with Pluto sitting right on her Neptune, which is a rare transit that can only happen once every 248 years, and this had been activating her Mars/Uranus/Neptune square for two years. Then along came Saturn which joined Pluto in a conjunction on her natal Neptune—and that very week she had the break, and landed in a mental hospital. The timing was uncannily precise, as was the accuracy of Ebertin's book. Of course, this example is just one anecdotal case, but it offered at least a partial answer to a painful conundrum I had carried in my heart for many years.

At one level this experience could be dismissed as a sheer coincidence. At another level, it seemed deeply significant. Saturn conjoining Pluto only occurs about every 36 years. So these rare Pluto and Saturn transits intersected and joined forces precisely on her sensitive Neptune aspect, which somehow connected her experience directly to something larger in the cosmos; and the character of that challenging moment was precisely indicated in the astrological texts. So it seemed that something larger was involved here, beyond just her own internal process.

I realized that I needed to explore these correlations further. This led to several years of tracking the outer planet transits of clients that I worked with, especially in holotropic breathwork. Some of these clients I worked with over a period of a few years, and I continued to observe astounding correlations, particularly between the transits of the outer four planets and the transformative experiences these people were having in holotropic breathwork, and also in psychedelic sessions.

In his decades of clinical research Grof frequently observed the "principle of synchronicity" in action, and his collaboration with Richard Tarnas amassed substantial evidence for astrology (Grof, 2009). Grof concluded that "astrology is in my opinion the only system that can successfully predict both the content and timing of experiences encountered in non-ordinary states of consciousness in experiential psychotherapy" (Grof, 1984).

What kind of universe makes synchronicity possible?

In 1990 the Dalai Lama gave a talk in which he said (paraphrasing): To discover the deepest secrets of the cosmos, there are four fundamental steps:

1. Explore the outer (visible) universe—*physis*
2. Explore the inner (invisible) cosmos—*psyche*
3. Explore the relationship between them.
4. From these three, one can discover the deepest secrets of reality.

Western science has generally focused primarily on step 1 above, often taking it to be the whole of reality. Anything beyond step 1 has generally met stiff resistance in Western science. For example, cosmologist Carl Sagan, borrowing from Marcello Truzzi, developed a cherished scientific mantra known as the Sagan standard: "Extraordinary claims require extraordinary evidence."

Fortunately, over the past 60 years, consciousness research has finally been providing extensive scientific evidence for step 2 in the Dalai Lama's schema above, which is the science of nonphysical realms of consciousness. As a result today, the demanded extraordinary evidence is now *in*—and this is presenting a major challenge to the prevailing mechanistic scientific materialist paradigm, which insists that nothing exists beyond the physical universe, meaning that only step 1 above is deemed legitimate science. Hence, the pioneering data from the frontiers of consciousness research are still widely ignored or denied in the scientific community today.

Yet, the tables have now turned: *extraordinary denial* of consciousness research demands *extraordinary refutation* of the voluminous evidence. The onus is now on the scientific materialist community to either refute or explain this evidence; and so far, neither is forthcoming. Failing this, the scientific materialist community must either acknowledge this extraordinary evidence as legitimate, including its far-reaching implications, or else find themselves in stark betrayal of the very scientific principles they purport to cherish and uphold.

Holomovement and Indra's Net

Regarding step 3 of the Dalai Lama's schema, physicist David Bohm was one of the earliest contemporary scientists to take consciousness seriously. Bohm was a quantum physicist of the highest caliber, and a close colleague of Albert Einstein. Bohm carried his quest for truth deeply within and *beyond* physics, and he engaged in extended dialogues with spiritual masters such as Krishnamurti, the Dalai Lama, and others.

In 1952 Bohm published a radical reformulation of quantum physics, based in part on nonlocality, which Einstein couldn't accept and famously dismissed as "spooky action at a distance." Bohm's breakthrough paper was a major wake up for physicist John Bell, who loved it except for the nonlocality, which he tried hard to get rid of. But, by Bell's own account, Bohm's quantum theory compelled him to realize that nonlocality is fundamental. This led Bell to develop his famous Bell's theorem, which in turn led to nearly 50 years of increasingly refined experimental tests, which consistently upheld both quantum mechanics and nonlocality. Three of the key experimentalists shared the 2022 Nobel prize in physics, and Einstein's dreaded "spooky action at a distance" is now confirmed to be real, and it forms the basis of today's pioneering "second quantum revolution." In my view, these experiments constitute remarkable scientific validation of the intrinsic *oneness* of the entire cosmos—something that mystics across the wisdom traditions have proclaimed for millennia.

Going beyond matter and energy as the fundamental scientific components of reality (and going beyond Einstein's $E = mc^2$), Bohm proposed back in the 1970s that there are *three* fundamental constituents of reality: matter, energy, and consciousness (or meaning), each of which enfolds the other two. This culminated in Bohm's theory of the implicate and explicate orders (corresponding to steps 2 and 1 respectively in the Dalai Lama's schema above), and also Bohm's elegant "holomovement" which proposes that the cosmos is "a single undivided wholeness, in flowing movement," in which "everything is connected to everything else" (Bohm, 1980).

This nonlocal interconnected cosmos is not Pollyannish wishful thinking; it has become the very cutting edge of quantum physics. Today it goes by the name quantum entanglement, although a more elegant and perhaps more qualitatively accurate term for it might be "quantum communion," or perhaps better, Thich Nhat Hanh's "interbeing."

Bohm's holomovement seamlessly weaves together the physical (matter-energy) and consciousness domains into an exquisite unified fractal holoverse (Bohm and Hiley, 2006). This constitutes Bohm's elegant answer to step 3 of the Dalai Lama's schema.

How does this all relate to synchronicity?

So now we finally come to the key questions: How could the planets or stars possibly have anything whatsoever to do with human affairs? The idea seems utterly preposterous! And what does all this have to do with synchronicity?

Everything! Bohm's holomovement, developed out of quantum physics, was recognized in an earlier form by ancient Hindu and Buddhist mystics, as expressed in the ancient metaphor of Indra's Net:

Reality is imaged as an infinite net of magnificent glittering jewels, each one of which reflects all the others. If you look into the facets of any one jewel, you see all the others reflected there. And if you look into any of the *reflected* jewels, you see all the other jewels reflected in that reflection. It's worlds of infinities contained within worlds of infinities, representing the fractal spiritual-material unity of cosmos and consciousness—at every scale from the infinitesimal to the Infinite. As the Indian scripture *Bhagavad Gita* expresses this stupendous realization (verse 6:29), the mystic "beholds the Self *(Atman)* present in all beings, and all beings in the Self."

When Indra's net is first encountered, people generally visualize the jewels as static in the net, and all of similar size. This is understandable (yet greatly oversimplified), as shown the figure below as a representation of Indra's Net.

An image of Indra's net, generated by AI graphic software. Notice how each jewel reflects all the others. This is a crude and static "snapshot" of a glorious unfolding cosmic evolution.

However, now imagine that we *animate* this exquisite lattice of jewels—in both space and time, which means we introduce both spatial and temporal variation. The jewels are now of all different sizes, ranging from tiny atoms to human-sized jewels to giant planet and star-sized jewels—all whirling around in grand dynamic motion. *Now* if you gaze into the facets of any one jewel, you will see all the glorious movements of this vast "universe" of swirling jewels, precisely replicated in dynamic miniature. Each jewel thus "contains" the entire cosmic dynamism, as it unfolds in holographic evolution on multiple scales of space, time, and consciousness simultaneously. This gives rise to the observed correlations across celestial and terrestrial domains and scales, all the way down to the individual human being, who is evidently a walking miniature cosmos.

I dare say this is one of the key secrets revealed in step 4 of the Dalai Lama's schema. Each of us has access to the full infinity of divinity, through our own hearts. This is how Meister Eckhart, for example, can proclaim that when the heart is purified, "God completely pours himself according to all his power; God holds nothing back in all his life, in all his being, in all his nature, in all his divinity" (Sermon 48, quoted in Hacket (2013).

In summary: Take Indra's Net, dip it into space-time, and you *necessarily* get an astrological hermeneutic. The explanation of planetary correlations with human consciousness then becomes straightforward: the planets (and the cosmos) are alive *in* us; they are intrinsically part of us, and we of them, as fractal interbeings—hence the observed correlations with planetary movements. *Whatever* the living consciousness of the planets actually is, that consciousness also dwells inside us as an integral part of *our* living being. Our physical eyes (and scientific instruments) are utterly blind to this, and deceive us into believing we are separate from the planets and the stars, simply because the planets are too far away and their physical forms are way too big to fit inside our physical bodies. But consciousness radically supercedes all that, in a living acausal fractal nonlocal *oneness*—across all levels of inter-being. In my view, this is a deeper implication and practical application of David Bohm's holomovement, which holographically interweaves matter, energy, and consciousness into an intricate, undivided, fractal wholeness at every level and scale of existence.

These are admittedly imaginal theoretical speculations at this point. Nevertheless, as Einstein said, "Imagination is more important than knowledge. For knowledge is limited, whereas imagination embraces the entire world, stimulating progress, giving birth to evolution. It is, strictly speaking, a real factor in scientific research" (Einstein, 1929).

References

Bohm, D. (1980). *Wholeness and the Implicate Order,* Routledge (London).

Bohm, D. & B. Hiley (2006). *The Undivided Universe: An Ontological Interpretation of Quantum Theory,* Routledge (London).

Einstein, A. (1929). Quoted in Viereck, G.S., "What Life Means to Einstein: An Interview," *The Saturday Evening Post,* October 26, 1929, p. 17, Quote Page 117, Column 1, Saturday Evening Post Society, Indianapolis, Indiana (verified on microfilm).

Grof, S. (1984). *Ancient Wisdom and Modern Science,* SUNY Press (Albany, NY).

Grof, S. (2009). "Holotropic Research and Archetypal Astrology," *Archai: The Journal of Archetypal Cosmology,* Vol. 1, No. 1.

Hackett, J. (2013). *A Companion to Meister Eckhart,* Brill (Boston), p. 350.

Chapter Four

Synchronicity in the Evolution of my Worldview

ELISABET SAHTOURIS, PHD

Elisabet Sahtouris, PhD is an internationally known evolution biologist, futurist and consultant. With a post-doctoral fellowship at the American Museum of Natural History, she taught at MIT, the University of Massachusetts, contributed to the NOVA-Horizon TV series, became a fellow of the World Business Academy with an honorary Chair in Living Economies, a fellow of the Findhorn Foundation, and is on the Advisory Board of Ethical Markets, Tree Sisters and other organizations. Dr. Sahtouris is a co-founder of the Worldwide Indigenous Science Network, convened international symposia on Foundations of World Sciences and taught at Chaminade University in Honolulu. Her most recent books are *Gaia's Dance: The Story of Earth & Us* and *VISTA: Life & Getting Where You Want to Be.*

My life has been a journey full of synchronicities, each of which has contributed to my evolving view of reality, from cosmological to mundane. Synchronicities have catalyzed important shifts and expansions in my worldview, from that of a materialist scientist, believing that the brain produces consciousness, to an individual who experiences consciousness in all aspects of the universe and the interconnection of all that exists. In short, synchronicities revealed the deeper reality behind our physical world in undeniable ways.

I will first describe a series of related synchronicities that occurred in different times and places, then another single synchronicity unrelated to those, and then my understanding of the nature of a universe that makes synchronicity possible. The first series gives a bit of the early evolution of my journey, and my move to a Greek island, where serial synchronicities occurred involving the Harmonic Convergence, the feathered serpent god, Quetzalcoatl, and a centipede. The second describes a synchronicity involving a magnificent manifestation soon after I arrived in Greece, exploring the ancient and magnificent theater and healing center complex of Epidaurus.

Synchronicity 1. Encounters with Quetzalcoatl

In the late 1960s, with a PhD in evolution biology and an NIH Post Doc at the American Museum of Natural History in New York City, I was still left with my biggest questions about the human condition unanswered. Into the 1970s, I read many intriguing books on cosmology and what was then called 'paraphysics', among them Erich Jantsch's *The Self-Organizing Universe,* David Bohm's *Causality and Chance in Modern Physics,* met David Peat, Bohm's co-author and later biographer, in Canada along with Leroy Little Bear and (Pam) Apela Colorado in our shared interest in indigenous science, which led to our forming the Worldwide Indigenous Science Network.

In the late 1970s, my children grown, and inspired by Henry Miller, whom I'd met in his California home where we had the most wonderful conversations, I left the US for Greece. Especially because of his book *The Colossus of Maroussi,* I intended to follow in his footsteps, writing novels to explain the human condition to myself. A decade later, in 1987, I had married a Greek fisherman and was living a simple life on a small, forested island. Having written a few novels, I decided to go back to science, writing many essays on philosophy, and, immersed so directly in Nature, to work out my own story of evolution biology.

Coming back to my island on the ferry from the larger island of Aigina, one day, I met an American tourist whom I helped to find a hotel. In gratitude, she gave me an amazing gift: a round-trip air ticket from Greece to Philadelphia with fixed dates that she could not use and which was not refundable. Though my financial resources were long dried up, I had no thoughts of going back to the US when suddenly this ticket dropped out of the blue by way of a stranger on a boat.

A day or two later, my phone rang—a rare occurrence, as it was only the 2nd phone to be installed in our tiny hamlet on a hill. It was a New York friend calling me from Athens, where she had flown on a sudden whim. When I told her about the ticket she asked for its dates and then I heard her say in surprise, "Oh! You must be meant to go to the Harmonica Virgins!" Wondering whether this was a concert by some feminist music group, I inquired who the Harmonica Virgins were. "No, no," she laughed, "the Harmonic Convergence, the fulfillment of the Mayan prophecy, the awakening of Quetzalcoatl!"

Now Quetzalcoatl was a name I could relate to—I'd known about the feathered serpent god since childhood, having steeped myself early in Mayan culture through avid reading. I was so into it I had even named my teddy bear Popocatepetl, after the Mayan volcano, loving the sound of the name.

No sooner had my friend spoken the god's name than goosebumps broke out all across my bare shoulders—I was shirtless in the summer heat, standing phone receiver in hand beneath the thatched roof of the old stone house I'd fixed up to make it livable. As she kept talking, the goosebumps did not go away, but seemed to move back and forth across my shoulders in waves. Brushing my hand over my tingling shoulder, off fell a huge centipede, squirming on the floor like a small-fringed snake. "Elaine!" I spouted out, "I'll go—the feathered serpent has just appeared in person!"

What a magical reality! The Harmonic Convergence, I learned, mostly after arriving at it in Philadelphia, had been called by a man named Jose Arguelles to celebrate an unusual configuration of planets on a significant Mayan Calendar date ending a long cycle of over a thousand years. It was about heralding a new world by creating a synchronized global peace meditation in various 'power center' locations, Philadelphia being considered one of the most important. Many interesting people had been brought together, including a number of indigenous leaders, and I gained several significant new friends there.

A highlight of the gathering for me was watching Chief Leon Shenandoah—Tadadaho (chief of chiefs) of the 'Six Nations' Haudenosaunee (whom the European settlers called Iroquois)—do a fire ceremony in a Philadelphia city park. As he prayed, a flock of wild geese flew over us, landed and walked circles around us. When I commented to Leon later how beautiful that was, he replied, "When you speak to Nature correctly, She always answers."

Oren Lyons, himself the Onondaga chief under him, was also there and we reminisced about being art students together at Syracuse University in the 1950s before I became a scientist and before he was appointed Indian chief. He was fascinated by some of the things I shared about scientific evidence for Earth as a living planet, especially my stories of our microbial ancestors. Oren pointed out that even though Indians know Earth as alive, it is good to be able to use microscopes and scientific research to reveal these important stories.

Further magic came into my life quickly after that, including three successive trips to Costa Rica. On the third, I visited the underground indigenous artifacts gold museum owned by a bank and approached by an elevator that took people beneath the central square of the capitol, San Jose. The atmosphere in the gold museum was dreamy, with dim lighting and small spotlights on the golden objects in glass cases, many of them representing spirit guardians to the ancient Tairona people who had made them.

Walking among them in a kind of reverie, I mentally addressed a query to Quetzalcoatl—whether he had ever appeared to his people as a humble centipede. No sooner had I formed the question in my mind than something caught my eye at the far end of the very large room, something glinting, moving. I walked straight to it and found a golden centipede swinging from a pair of threads inside a glass case. Nothing else moved, and as I continued through the museum, I found no other such image, nor anything in motion.

The live centipede falling onto and from my shoulder to signal the Harmonic Convergence and the swinging gold one in the case seemed meaningful messages coming to me—not as visions, but as real, if surprising, events in the physical world. They seemed connected to my growing work with indigenous people—people who had not lost their natural relationships within all of nature—their richly intelligent world apparently still spoke to anyone who would acknowledge it in gratitude and listen. I scoured dusty shops for centipede images, finally finding one on an old ceramic plate in Mexico—after all, the birthplace of Quetzalcoatl.

Synchronicity 2. The manifestation of a bank on wheels

Setting out on my own from Athens to Epidaurus, for my first visit to the great amphitheater and whatever else remained of this ancient healing center complex, I shunned package tours to work the trip out on my own, starting with a boat from Piraeus, the port of Athens, to Old Epidaurus on the coast, where I stayed overnight in a small family hotel to take a very early morning local bus up the long winding road to the ruins.

The driver dropped me off at the nearest point to the theater, indicating the direction in which I was to walk, though the thick early morning fog made it hard to see the road ahead at all. Only as the mist gradually lifted along what I later calculated as a full five kilometers to the gate, did fields carpeted in blood-red poppies beneath olive trees reveal themselves in all directions. Tears of joy ran down my cheeks at their sheer beauty as my heart overflowed in immense gratitude for being privileged to walk this ancient sacred ground.

When I finally arrived at the magnificent theater—the best preserved in all Greece—I was also grateful to have it all to myself, clambering over the tiers of stone benches, imagining all the olive pits that must have piled up during the day-long therapeutic performances, generally both a comedy and a tragedy on the playbill—talk about wide perspectives!

The theater's awesome acoustics made a pebble dropped on the stage floor audible in the great arc of its highest seats. Thus, some ten to twenty thousand people could gather in one place to learn about the human condition from highly entertaining actors speaking on a distant stage with no sound equipment other than a resonating tank of water buried underneath that stage and the open architecture of the theater itself.

Trekking through the rest of the ruins at Epidaurus, I gradually got a sense of what a magnificent spa this had been in its heyday. Amazing that these sun-whitened, scattered ruins of temples and hotels could still convey the original splendor of the grandness once all colorfully painted, just as in Egypt—seemingly endless stone-carved friezes of lion heads and flowers, fountains, statues and chariot racetracks, the gaiety of comedy played alongside tragedy for their deep psychological lessons about the actions of gods and mortals in a layered, or fractal, universe where everything affected everything else.

It seemed so clearly a celebration of life. No wonder people healed here—it was about so much more than sleeping with snakes on stone temple floors to inspire healing dreams. Well-being still seemed to ooze from these stones—bleached bones of their glorious past yet speaking to me of their original beauty. No wonder Henry Miller had sent postcards from here to three doctor and psychiatrist friends, telling them to come themselves, adding the biblical quote: *Physician heal thyself!* and then laughing when one proved to be deceased and the others unresponsive.

Preparing to continue from Epidaurus to Nauplion by bus for my next stop on this intended three-day tour, I went to the government pavilion and hotel next to the ruins intending to cash a Travelers check, only to find there was no place to do that, no way at all to get any cash. Banks had not been open before I'd set out from Piraeus the day before and I had only a few coins left in my pocket, not nearly enough even for the Nauplion bus.

Completely stymied by this unforeseen dilemma, with no means to go on and nowhere to sleep, I went into the hotel restaurant to think and spent my last coins on a Greek brandy, the only thing on the menu I could cover. As

I sipped it, an agitated waiter suddenly ran up to me, grabbed my arm and pulled me rather roughly from my chair. I had no idea what I had done to offend him and get myself thrown out. He dragged me to the door and pointed into the road, as though to yell "Get lost!" at me.

Then I saw it—an official-looking vehicle like an RV with a side door, a step, and a clerk inside. Apparently, a little bank on wheels had pulled up! Sure enough, the clerk cashed my travelers check with a smile and I stepped back out into the road still dazed by the cash in my hands as I watched the vehicle pull away again without having attracted the attention of anyone else. Even the waiter still looked shocked himself as I regained the presence to thank him—at least I'd learned *that* word—and silently thanked Hermes the Trickster, the god of travels, for this magical good fortune.

In all my years in Greece when I returned to live there, I never saw another such bank on wheels. I did continue a habit begun after the incident at Epidaurus, of making daily gratitude offerings to Hermes, in the form of small coins given to beggars encountered wherever I went, and I did have many other delightful anomalous experiences after that one (Elisabet Sahtouris, 2023, pp17,18).

What kind of universe makes synchronicity possible?

To develop a more accurate and inspiring scientific model or worldview as a framework for the human journey, one that makes synchronicity possible, we have to recognize the whole universe, Earth within it, and humanity within Earth, as a coherent living system with system dynamics. Very helpful to this Big Picture thinking is the recognition that nothing is independent of context and that context is what gives meaning to things, to phenomena. This I derived from Arthur Koestler's concept of holons in holarchy, a holon being any entity, and its contexts being the larger holons in which it is nested. In nature, we find particles exist in atoms, atoms in molecules, then on into cells, organs, bodies as a holarchy that continues with families, ecosystems and on to Earth, solar system, galaxy all the way to the universe. Everything is embedded in holarchy and holarchies intersect. It is this interconnected universe that underlies the phenomenon of synchronicity, and was a key to my understanding both of my synchronicities, described above.

Another vital concept of a universe that makes synchronicity possible is that consciousness is primary; a replacement for the existing scientific belief in the foundational concept of things the other way around: that matter is primary and consciousness a late emergent product of its biological evolution. Harvard's Nobel Laureate biologist, George Wald, in his essay, 'Life and Mind in the Universe (1984), spelled his own view of consciousness out this way:

> *Consciousness is not part of that universe of space and time, of observable and measurable quantities, that is amenable to scientific investigation. For a scientist, it would be a relief to dismiss it as unreal or irrelevant. I have heard distinguished scientists do both... Unfortunately for such attitudes, consciousness is not just an epiphenomenon, a strange concomitant of our neural activity that we project onto physical reality. On the contrary, all that we know, including all our science, is in our consciousness. It is part, not of the superstructure, but of the foundations. No consciousness, no science. Perhaps, indeed, no consciousness, no reality.*

A third vital concept for a cosmos in which synchronicities happen is that deep reality (beyond the physical) has no timeline, and is thus an 'eternal Now. Indeed, even in a physical world where timelines must be set up for us to function, each of us only experiences a constant Now, in which we can conjure past and future, but always only Now. Amazing to me that the science in which I was trained completely ignores this seminal fact, along with the related fact that all experience lies within consciousness.

My favorite metaphor for the entre cosmos, as well as for any living being within it, is an endless keyboard of vibrations in which matter occupies the low keys, energy continues along the midrange and mind, spirit, consciousness the upper ranges. We are not bodies with minds and spirits or souls, but are the entire range seamlessly integrated (Sahtouris, 2022).

Synchronicities imply that our minds have no boundaries and so cannot be contained in time or space. If we conceive the universe itself as alive, everything evolving naturally within it is alive and therefore self-transcendent, consciously knowing itself and thus knowing the whole at the fundamental level of cosmic consciousness.

Recently, I listened to a talk by Ibrahim Karim on his cosmology, speaking of two kinds of perception of 'reality': multidimensional and linear, the latter being the limited perception of our physical timeline reality. He then spoke of the hybrid version in which both kinds of perception are available. For example, some individual humans are born multidimensional and so have to learn the linear mode. It took one such individual, a dear friend, forty years to be able to see a wall as a flat surface rather than a hologram. Everything in Nature other than humans is in multidimensional mode, which is why such individuals find it easier to be with animals, trees, etc. Only in linear mode is there a timeline, analysis, theorizing, fear of death and so on and on. I believe this helps explain synchronicities as multidimensional events showing up in our linear world.

And of course, it is important to realize that if the living universe story is true, then our cells are conscious, intelligent beings in the great community of our bodies, and every mushroom and mongoose knows its place in its living community. Everything is connected. Looking at all nature from this foundational conceptualization changes everything. The world is a cooperative living community, in which synchronicities occur naturally. As I wrote in my book with Willis Harman (1998):

> *The new biology would show us that mature ecosystems do not evolve toward the sole survival of the most aggressive and clever, as they sometimes do in early stages of self-organization or when invaded by very aggressive 'outsider' species. Rather, they evolve toward intelligent and cooperative mutual support communities in which every species has a valid and valued role. In such a community, the bottom line is not profits, but output useful to other species in a virtually 100% recycling economy.*

This interconnected cooperative mutual support is the basis of a universe that supports the spontaneous occurrence of synchronicity.

References

Harman, W., & Sahtouris, E. (1998). *Biology Revisioned.* North Atlantic Books

Sahtouris, E. (2022). A Keyboard Metaphor, *Paradigm Explorer,* Scientific & Medical Network, vol 2.

Sahtouris, E. (2023). *VISTA: Life & Getting Where You Want to Be,* pp17,18. Upaya House.

Wald, G. (1984). Life and mind in the universe. *Quantum Chemistry* Supplement: Proceedings of the International Symposium on Quantum Biology and Quantum Pharmacology, 26, S11: 1–15.

Chapter Five

Synchronicities: From the Sublime to the Ridiculous

DEAN RADIN, PHD

Dean Radin, PhD is Chief Scientist at the Institute of Noetic Sciences (IONS), Associated Distinguished Professor at the California Institute of Integral Studies, and chairman of the genetic neuroengineering company, Cognigenics Inc. He earned an MS in electrical engineering and a PhD in psychology from the University of Illinois, Urbana-Champaign. Before joining IONS in 2001, Radin worked at AT&T Bell Labs, Princeton University, and the University of Edinburgh. He has given over 750 invited talks and interviews worldwide, and published some 300 articles, four dozen book chapters, and four best-selling books translated into 15 foreign languages.

Introduction

Synchronicities involve meaningful correlations between the mental and material world (Atmanspacher, 2018). A mundane explanation for such events is that with some 8 billion people in the world having perhaps a thousand unique experiences per day, some of the resulting 8 trillion experiences will include remarkable coincidences purely by chance. From the perspective of the individual who experiences one of those coincidences, it would seem exceedingly unlikely to just be due to chance, so it must be meaningful. However, when dealing with such a large sample size even a billion to one outcome would not be unexpected. Someone will always win the lottery purely by chance. Still, such coincidental events are so awe inspiring that tales will be collected and interpreted as evidence of something from beyond, often cast in terms of spiritual or divine guidance (Beitman, 2022; Rushnell, 2018; Schwartz, 2015).

To test if pure chance is the only possible explanation for synchronicities, since the 1930s researchers have conducted mind-matter experiments in the laboratory where chance outcomes are controlled and established in advance.

Experiments using tossed dice (Radin & Ferrari, 1991), true random number generators (Radin & Nelson, 1989), and dozens of other physical targets (Heath, 2011; Jahn & Dunne, 1987; Radin et al., 2013; Shiah et al., 2017; Shiah & Radin, 2013), all indicate that mind can influence matter beyond chance. This means that some synchronicities are "real" in the sense that somehow mind, matter, and meaning can coalesce into remarkable coincidences, in the literal sense of planned outcomes occur at the same time. This does not mean that everything labeled a synchronicity is necessarily something other than chance, but it does mean that in principle it might be.

A philosophical perspective

How can we understand these experiences? Dual-aspect monism (DAM) is a philosophical approach that offers some ideas (Atmanspacher & Rickles, 2022). DAM proposes that mind and matter are epistemic aspects of an underlying holistic, "psychophysical" realm. Carl Jung called this primordial reality the *unus mundus,* the "one world" (van Erkelens, 1990). Similar concepts can be found across the world's esoteric traditions. Within DAM, the undifferentiated reality of the *unus mundus* is said to split into different aspects, two of which we may experience as mind and matter.

The split occurs via *meaning*, which posits a special kind of relationship between mind and matter. The word "meaning" commonly refers to a felt sense of special significance related to a symbol or a word, and in the present context meaning may be thought of as a mediator that connects subjective experiences with the physical world, and vice versa. That is, it is a relationship between mind and matter. Under special circumstances, when this relationship is especially vibrant, or imbued with unusually intense need, the correlation between mind and matter becomes tightly bound, metaphorically like two sides of the same coin.

According to DAM, mind and matter also react back into the *unus mundus,* setting up a recursive relationship. Mind-matter correlations do not involve ordinary causal forces or interactions in the sense of Aristotle's "efficient cause." Rather, mind and matter are said to be related to each other *acausally,* analogous to quantum entanglement, where strong correlations are observed between isolated entities (like photons or electrons), but without causal links between them.

To help contrast DAM with other philosophies, it is useful to briefly review the three primary contenders. First, *dualism* proposes that mind and matter are two fundamentally different "substances." The traditional complaint about dualism is that if you have two radically different substances, how can they possibly interact? We know that they do interact because if you drink an entire bottle of wine, or drink five triple expressos, there's no question that both your mental and physical states will change in predictable ways.

Second, *idealism* proposes that consciousness is fundamental. In this case, the physical world is regarded as a mere appearance, i.e., a mental inference. The problem with idealism, as Einstein once put it, is "Is the moon is still there when nobody looks?" (Mermin, 1985). That is, the physical world certainly seems to have its own stable existence and set of rules, independent of us, so idealism requires that we attribute the appearance of the physical world to a cosmic or universal consciousness. While some might dismiss such a possibility, there are fascinating arguments in favor of this concept (Kastrup, 2021).

Third, *materialism* is the prevailing philosophical assumption within the sciences today. It is adopted, often unknowingly, as the only reasonable way to think about the nature of reality, largely because of its undeniable success in creating new technologies. Materialism is the flip side of idealism, in that it proposes that everything, including consciousness, is ultimately physical. As Francis Crick put it, this philosophy asserts that you, your thoughts, your dreams, and your subjective sense of identity, are "nothing but a pack of neurons" (Crick, 1995). The problem with materialism is that no one has any idea about how the subjective taste of an orange, the first person experience of the color red, or what it is like to be *you* from the inside, can emerge from a physical substrate. As philosopher Jerry Fodor put it in 1992, and which is still true today, "Nobody has the slightest idea how anything material could be conscious. Nobody even knows what it would be like to have the slightest idea about how anything material could be conscious" (Fodor, 1992).

Given the limitations of the above three philosophies, psychiatrist Carl Jung, prominent physicists Wolfgang Pauli, Arthur Eddington, John Wheeler, David Bohm, and others, proposed variations of DAM based on considerations motivated by quantum theory. They considered mind and matter to be in a deeply connected but acausal relationship that was not localized in space or time, and they all noted the key role of *meaning* in this relationship (Atmanspacher & Rickles, 2022).

The strength of DAM is that it offers a framework for understanding synchronicities. For example, consider an experiment seeking to demonstrate, say, a telepathic connection between an isolated pair of people. If the telepathic receiver successfully describes a randomly selected image that a distant sender is attempting to mentally share, that outcome is considered a meaningful "hit" rather than mere coincidence, because the experimental protocol *assigned meaning* to that particular type of outcome. In this way, an experimental protocol transforms an otherwise chance event into a meaningful, non-chance event (where scientific confidence in that interpretation is strengthened by repeatability, which is also part of the protocol).

Sublime synchronicities

The substantial experimental evidence in favor of the existence of psychic phenomena (psi, for short) supports the idea that psi experiences, whether evoked in the lab or spontaneously experienced in real life, are forms of synchronicity (Cardeña, 2015; Radin, 2018). I propose that the experimental studies of psi are a *sublime* form of synchronicity, because in such cases we can exclude mere coincidence as an explanation. This establishes synchronicity as a much more interesting phenomenon than, say, dumb luck.

While psi experiments are persuasive, they require an appreciation of research methodologies and statistical methods to fully grasp their impact. Fortunately, there are other types of synchronicities that I call *ridiculous*, not in the sense of silly, but in the sense that the experience requires an outrageous stretching of the imagination. I have had a number of preposterous synchronicities. Here is an example.

The spoon episode

Around 2000, I attended a "spoon-bending party," which was a lighthearted entertainment offered as an evening event at a conference I was attending. At such events, attendees attempt to psychically bend metal spoons or forks. I had heard about such events for years, and some of my friends and colleagues who had attended such a party later proudly showed me their bent or twisted spoons. But this upcoming event was the first time I had a chance to personally participate, and even though I trusted that my friends were not telling me tall tales, I was very skeptical that the reported effects could be anything other than dissociated feats of strength. It is easy to bend the neck of a metal spoon or fork.

At the event, I was determined to watch someone in the group do the impossible—bend the bowl of a large, heavy soup spoon in half. I wanted to see this happen right in front of me. I chose this as my goal because it is virtually impossible to bend the bowl of a soup spoon without inflicting serious damage to your fingers, and even nontrivial to accomplish if you're using a metal gimmick to provide leverage.

At this event, we were told that if we were able to bend the bowl of a heavy soup spoon with a gentle touch, we would be awarded with a big yellow button to wear, which had the words "warm former" printed on it. This strange term referred to the idea that somehow the mind would make the metal warm, and then you could easily form it into shapes. People who said they could do this typically reported that the metal momentary felt like putty, which was then easy for them to move into other shapes.

For reasons I do not understand, when I learned that I could get a big yellow button, I desperately wanted, no—I *needed* to get one of those buttons, because the next day I would be giving a talk at the conference, and I saw myself at the podium wearing the yellow button. I normally shy away from any form of self-aggrandizement, but at that moment the thought of winning a yellow button was strangely alluring. To be clear, this feeling wasn't an idle ego-driven motivation. It was a crazed obsession. I felt that the fate of the universe depended on my getting that button. I remember this odd state very clearly because it was so unusual for me.

In any case, that manic moment faded into the background a bit after I found a woman at the party who claimed that she had bent the bowl of a spoon in a previous party. I stood about 18 inches in front of her and closely mimicked what she was doing with her spoon. She held the stem in her right hand and the bowl pinched lightly between her left thumb and forefinger.

After a few minutes nothing was happening, and I was getting tired because my full attention was on her spoon, and I was tightly holding on to my spoon. Then suddenly my attention was distracted for a moment by someone in the room shouting with delight, signaling that they had probably bent a spoon or fork. A second later, someone behind me yelled, "Look what you've done!" I looked around to see what the commotion was, and the person said, "No, look what *you've* done!" Meaning, me. I looked, and to my astonishment I had bent the bowl of my spoon about 90 degrees. I immediately examined my fingers to see if there were indentations or other signs of force on my fingers. There weren't. So I poked the bowl of my spoon with a finger, and it felt soft like putty, just like others had reported. I pinched it, and the bowl easily folded all the way over. This is a photo of that spoon.

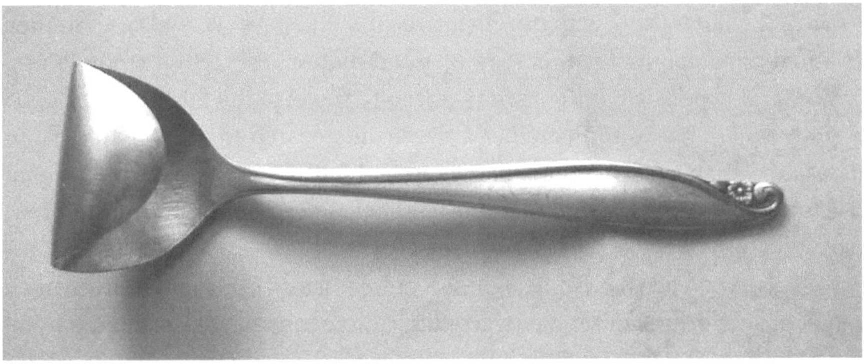

A second later it stiffened up and it has remained that way ever since. It was cool to the touch throughout this event. Later, I checked to see if it was a real metal spoon. It was indeed. I determined that it wasn't nitinol (also called memory metal), which can adopt a different shape if subjected to heat (and is a common gimmick that magicians use to fake metal-bending). It doesn't change shape when placed in boiling water or iced water. The bend itself is entirely smooth, unlike what a spoon bent by force would typically look like. Using a torque wrench, I measured the amount of force required to make that bend in a similar spoon from the same manufacturer, and it was far beyond what I, and I'd guess anyone, could accomplish using only two fingers.

The day after this party I gave my talk on stage while wearing that big yellow button, and while doing so I felt that the universe had returned to its proper state, and the crazed sense of motivation that I had experienced was gone.

I regard this experience as a ridiculously impactful synchronicity because it involved a mind-matter interaction with an unusually intense injection of meaning. In retrospect, if I had not experienced this first-hand, I would have remained very skeptical that something like this was even possible. This synchronicity was thus transformative above and beyond the many successful psi experiments I have conducted, because this episode significantly revised my sense of the limits of what the mind and meaning can do.

A funding shortfall

This is a second example of a thoroughly ridiculous synchronicity. In early 2023, I was awarded a grant to experimentally investigate the quantum observer effect. I had previously conducted about two dozen studies on this topic in the laboratory, with promising results (Radin et al., 2021). But finding talented participants for this type of mind-matter interaction study is non-trivial, and expensive, if they are required to travel to the lab. So, I designed a study where we would conduct a worldwide search for 50 suitable candidates, and then mail them a suitable optical interference device they could use to perform the experiment in the comfort of their home, and at their own pace.

This required that we design and build custom optical systems, which required expertise in opto-electronics, microcontrollers, printed circuit board design, electronics fabrication, and computer programming. My grant proposal assumed that one of our staff members, who had the requisite expertise, would provide the necessary work and equipment at no cost. That was an exceptionally gracious offer which allowed the grant to be funded, because otherwise the proposed budget would have been too high.

Just as the grant was awarded, my colleague became ill and unfortunately could no longer build or fund the required equipment. Now I was in a panic because there wasn't enough in the budget to hire someone to make the devices as specified in the grant proposal. I was faced with one of two choices: either return the grant funds or find an additional $25,000. I did not relish the thought of returning a grant, so I set my intention to somehow resolve this problem in a more favorable way. My motivational state was similar to that described in the spoon episode. I *needed* a viable solution, and fast.

One week later I received an email from a new organization interested in esoteric practices, including traditional magical techniques, and they invited me to be on their advisory board. One of their initiatives would be to sponsor a scientific test of a magical practice (i.e., the real kind, not stage magic). I asked how much that grant would be. The answer, which you can probably guess, was €25,000. In terms of dollars, this was just over $25,000. And yes, I got that grant.

Conclusion

Based on a half-dozen outrageously synchronistic episodes like these, I have come to agree with Willis Harman, president of the Institute of Noetic Sciences from 1978 to 1997, who said, "Perhaps the only limits to the human mind are those we believe in" (Harman, 1998). And when aiming to manifest an extremely unlikely synchronicity, apparently it also helps to attain a peculiarly crazed motivational state.

References

Atmanspacher, H. (2018). Synchronicity and the experience of psychophysical correlations. In *Research in Analytical Psychology* (pp. 227–243). Routledge.

Atmanspacher, H., & Rickles, D. (2022). *Dual-Aspect Monism and the Deep Structure of Meaning.* Routledge.

Beitman, B. (2022). *Meaningful Coincidences: How and Why Synchronicity and Serendipity Happen.* Park Street Press.

Cardeña, E. (2015). *Parapsychology: A Handbook for the 21st Century* (J. Palmer & D. Marcusson-Clavertz, Eds.). McFarland.

Crick, F. (1995). *The Astonishing Hypothesis: The Scientific Search for the Soul.* (Reprint edition). Scribner.

Fodor, J. (1992, July 3). The big idea: Can there be a science of mind? *The Times Literary Supplement*, 5–7.

Fodor, J. (2006). How the Mind Works: What We Still Don't Know. *Daedalus, 135*(3), 86–94.

Harman, W. (1998). *Global Mind Change: The Promise of the 21st Century.* Berrett-Koehler Publishers; Second edition.

Heath, P. R. (2011). *Mind-Matter Interaction: A Review of Historical Reports, Theory and Research.* McFarland.

Jahn, R. G., & Dunne, B. J. (1987). *Margins of Reality.* Jovanovich.

Kastrup, B. (2021). Analytic idealism and psi: How a more tenable metaphysics neutralizes a physicalist taboo. In E. F. Kelly & P. Marshall (Eds.), *Consciousness Unbound: Liberating Mind from the Tyranny of Materialism* (pp. 257–283). Rowman & Littlefield Publishers.

Mermin, D. N. (1985). Is the moon there when nobody looks? Reality and the quantum theory. *Physics Today*, 38–47.

Radin, D. (2018). *Real Magic: Ancient Wisdom, Modern Science, and a Guide to the Secret Power of the Universe* (1st Edition edition). Harmony.

Radin, D. I., & Ferrari, D. C. (1991). Effects of consciousness on the fall of dice: A meta-analysis. *Journal of Scientific Exploration,* 5(1), 61–83.

Radin, D. I., & Nelson, R. D. (1989). Evidence for consciousness-related anomalies in random physical systems. *Foundations of Physics,* 19(12), 1499–1514. https://doi.org/10.1007/BF00732509

Radin, D., Johnston, J., & Delorme, A. (2013). Psychophysical interactions with a double-slit interference pattern. *Physics Essays,* 26(4), 553–566.

Radin, D., Wahbeh, H., Michel, L., & Delorme, A. (2021). Psychophysical interactions with a double-slit interference pattern: Exploratory evidence of a causal influence. *Physics Essays,* 34, 79–88.

Rushnell, Sq. (2018). *When God Winks: How the Power of Coincidence Guides Your Life.* Howard Books.

Schwartz, G. E. (2015). God, synchronicity, and postmaterialist psychology II: Replication and extension of real-life evidence. *Spirituality in Clinical Practice,* 2(1), 86.

Shiah, Y.-J., Hsieh, H.-L., Chen, H.-J., & Radin, D. I. (2017). Effects of Intentionally Treated Water on Growth of Arabidopsis thaliana Seeds with Cryptochrome Mutations. *Explore, 13*(6), 371–378. https://doi.org/10.1016/j.explore.2017.05.001

Shiah, Y.-J., & Radin, D. (2013). Metaphysics of the Tea Ceremony: A Randomized Trial Investigating the Roles of Intention and Belief on Mood While Drinking Tea. *Explore, 9*(6), 355–360. https://doi.org/10.1016/j.explore.2013.08.005

van Erkelens, H. (1990). The unus mundus (One World) as meeting ground of science and religion. In J. Fennema & I. Paul (Eds.), *Science and Religion: One world—Changing perspectives on reality* (pp. 201–201). Springer Netherlands. https://doi.org/10.1007/978-94-009-2021-7_21

Chapter Six

Mystery in the Chapel

RICHARD TARNAS, PHD

Richard Tarnas is Professor Emeritus at the California Institute of Integral Studies, where he founded the graduate program in Philosophy, Cosmology, and Consciousness. He has taught courses in the history of ideas, depth psychology, archetypal cosmology, cultural history, and the evolution of consciousness. He has also frequently lectured on archetypal studies and depth psychology at Pacifica Graduate Institute, and was formerly the director of programs and education at Esalen Institute. He is the author of *The Passion of the Western Mind,* a history of Western thought widely used in universities, and *Cosmos and Psyche,* which received the Book of the Year Prize from the Scientific and Medical Network and contains an exploration of the phenomenon of synchronicity.

The highway is for gamblers, better use your sense
Take what you have gathered from coincidence
Bob Dylan

In February 2000 I learned that my dear friend Charles Harvey had died in England, after suffering for several months from an aggressive cancer. He was not yet sixty. His death was a deep loss for many of us. Charles was an uncommonly generous and learned man, the president of the British Astrological Association and a philosophical Platonist. There was a funeral soon afterwards near his home in Somerset which I was not able to attend, but at which a rainbow suddenly appeared as people left the church in Mells, remaining in the sky throughout their drive to the cremation service in Bath. That summer a larger public memorial service was held at St. James's Church Piccadilly in London, where I delivered a tribute.

Several of Charles's extended family members lived in the San Francisco Bay Area, where I live, and they wished to honor him and his life here in California with an informal private ritual. Not long after his death, about eight or ten of us made plans to do so. One of the family suggested that we meet on a Sunday early afternoon in the Presidio of San Francisco at the Social Club, where she thought there might be a good place to have our gathering. Unfortunately, it was apparent as soon as we all arrived there that this location would not work, being crowded and busy with socializing people and not offering a quiet space appropriate for our ritual. Quickly searching my mind for an alternative location, I remembered there was a chapel not far away within the Presidio grounds, on a hill overlooking the San Francisco Bay, which I had seen from a distance on other visits to the area. So on the off-chance it would be open and empty, we all drove there and walked in just as a minister was about to close the chapel, the Sunday morning service having finished some time earlier. I described what we were hoping to do, and he kindly allowed us to come in for that purpose. We all walked into the chapel and made our way to the front three or four pews on the right where we all sat down.

The first member of the family who was going to offer her remembrance of Charles stood up and walked to the front, turned around, and began to speak to us. But within the first couple of minutes, she suddenly stopped speaking and looked pale and amazed, as if she were seeing a ghost. She was looking just over our heads to the wall immediately adjacent to where we were sitting. She pointed her finger at a bronze plaque on the wall. There, among several other such plaques, each one honoring a veteran soldier from the past who was buried at the Presidio, was a plaque which read "In memory of Colonel Charles Harvey, 1875-1962."

I can tell you that the feeling of the uncanny was vividly present for all of us, so astonishing was the coincidence. We had not planned in advance to meet at that place, none of us had ever been inside the chapel, the entire situation had spontaneously unfolded in the moment. Our choice of that part of the chapel to sit in was quite by chance. And yet there in this chapel, as if waiting for us to gather beside it, was a bronze plaque already created and installed that solemnly commemorated Charles Harvey.

Reflection

Synchronicities seem to constellate with greater frequency and intensity of significance at threshold periods in life—births, deaths, existential crises, falling in love, spiritual emergencies and awakenings. Clearly this one fit that general category. But in addition, those close to a recently deceased person often report sensing their posthumous presence, communicated in an extremely wide variety of ways. All of us attending the memorial that day clearly felt that our friend Charles was present among us in some mysterious but very real way.

Yet what is also striking was the sheer creativity of the synchronicity, seemingly improvised by the universe on a moment's notice. One can imagine all sorts of ways this might have come about through some invisible orchestrating intelligence, or perhaps spirits, planting the general idea of meeting at the Presidio in the first place, then leading us gradually to the chapel and to the exact spot where the ritual could take place in sacred space. Or perhaps this synchronicity simply reflected the profound artistry of life at work, continuously unfolding with a kind of improvisatory genius. An additional artistic touch was the year of Col. Charles Harvey's birth, 1875, which is also the year of birth for C. G. Jung, who of course first formulated the concept of synchronicity and coined the term.

The contributors to this volume have witnessed an enormous range of extraordinary and memorable synchronicities. We all have reached the conviction that there are many more things in heaven and earth than are dreamt of by the reductionist materialism characteristic of the mainstream modern mind-set. For those who have attuned their ears to hear and eyes to see, opened their minds and hearts to the subtle ongoing reality of synchronistic phenomena around them, events like this intimate a larger intelligent order at work in life—deeper purposes, boundless creativity, and above all, a sense of numinous mystery that underlies all our lives.

Chapter Seven

Synchronicities as Examples of Universal Connectivity Beyond Space and Time, and Access to a State of Open-Eyed Meditative Flow

MARJORIE WOOLLACOTT, PHD

Prof. **Marjorie Woollacott,** Ph.D., is a member of the Institute of Neuroscience at the University of Oregon, and prior chair of the Dept. of Human Physiology. She is President of the Academy for the Advancement of Post-Materialist Sciences, on the Science Committee of the Institute of Noetic Sciences (IONS), Co-Director of the Galileo Commission, and Research Director for the International Association of Near-Death Studies (IANDS). Woollacott has published over 200 scientific articles and written or edited nine books, including *Infinite Awareness: The Awakening of a Scientific Mind* (receiving 8 book awards) and *Spiritual Awakenings: Scientists and Academics Describe Their Experiences.*

What is the nature of synchronicity? I believe each instance may display its own signature set of signs, with the nuanced meaning being different for each individual, depending on their lens on reality and the lessons they have set out to learn in this life. In this exploration of the nature of synchronicities, I will illustrate key aspects of three experiences that awakened me to a broader understanding of the nature of reality, a clue to the significance of synchronicity for me. These insights are that in certain circumstances, we are able to become aware, 1) of aspects of consciousness beyond the normal space/time continuum, 2) that all conscious entities are connected or entangled, and 3) that there appears to be a state, that we might call 'flow', where we connect with this universal creative intelligence, and synchronicities occur effortlessly, in very powerful ways.

Synchronicity 1: September, 1987—I had spent about half of my sabbatical year—my 7th year at the University of Oregon—as a staff member, with my meditation teacher, offering service for her various programs, as she visited various meditation communities around the world. At the end of the sabbatical year, about a week before I was to return to Eugene, and my position at the U. of Oregon, she came into the office where I was working, and sat down in her chair.

She then turned to me and said, "So, Marjorie, are you getting married?" I thought that was an unusual question, as I had a boyfriend, but not someone I was at all interested in marrying. And so I said, "No, I'm not interested in getting married, because I couldn't travel with you as I did this year, if I were married." She didn't say anything more to me and began talking to someone else.

About a week after I returned home, I received a call from someone who had heard of my meditation path and was interested in coming to a meditation program at my home, which was the local meditation center. We talked on the phone for a while, and he was intrigued, and came to the next program about a week later.

When he walked in the door, and I saw him, I had the sense I was going to marry him.

Of course, I didn't say anything. About seven weeks later, he stayed after everyone else had left, and said, "I have something to confess to you." I said, "What?" And he said, "I am incredibly attracted to you!" I laughed, and said, "It took you long enough!" And we were married the following summer, and have now been married 35 years.

Personal significance: This synchronicity between my meditation teacher's question, and my immediately meeting my future husband immediately after returning home, deeply affected me. I tell this story again and again when people asked how my husband, Paul, and I met. It is important to me partly because it began a new phase of my life, of being lovingly yoked to another person whose primary goal in life was to experience a wider consciousness - and to experience it every moment. It was also one of the first times I was aware of an event in which time and space weren't operating in their usual ways of 3-dimensional reality.

Synchronicity 2. June 6–7, 2023—I had not been able to walk in the national forest behind my house in Sedona for a year, after falling and injuring my hip, and thus have been walking on even terrain on the roads near my house. My sister told me on Tuesday that she had met one of my old friends, Don, who I would often chat with when I met him on my hikes in the national forest—as she hiked in the forest that morning, and he asked how I was doing. I had a strong desire to see him again, and considered trying to go out to the forest the next morning, in case I might come across him but it didn't seem practical.

The next day, Wednesday, I delayed my usual walk on the road slightly, as I was packing to leave for a trip later in the morning. As I walked down the road, who should I see, but my friend Don, who was walking up the road with his dog, toward the national forest. We hadn't seen each other in a year, though I walk on that road near my house almost every day during my walks. We had a wonderful "reunion" and exchanged phone numbers so that we could keep in touch.

Personal significance: As this happened very recently, the significance was one of a joyful confirmation that the wishes of the heart are heard somehow within the nonlocal web of reality—and an answer is created that re-arranges events so that the wish of the heart is fulfilled.

Synchronicity 3. October 2015–June 2023—A series of events which seemed highly unlikely, and which led to my being invited to co-create the AAPS with Gary Schwartz, and later becoming president of the AAPS, and also being invited to be Co-chair of the Galileo Commission with David Lorimer.

In 2015, after being a professor of Neuroscience and Rehabilitation Medicine at the University of Oregon during the majority of my career, I published a book, *Infinite Awareness: The Awakening of a Scientific Mind* (2015), which was about my spiritual awakening and my subsequent exploration of the nature of consciousness. When I was working with the publicist for the book, she asked me what I hoped for from this publication, and I told her that I probably couldn't expect a best seller, but I would like to be able to speak at about four conferences a year on the topic of the fundamental nature of consciousness. At this point I didn't know anyone in the field, beyond having read their research papers as I wrote my book—and contacting, and then meeting, Ed Kelly and his colleagues at the University of Virginia, toward the end of the writing process.

As I moved forward, what seemed like synchronicity after synchronicity began to occur in my life. For example, I gave a first presentation of the topics of my book at the IANDS (Intl. Assn. of Near-Death Studies) conference in 2016, related to spiritual awakenings and NDEs, and really enjoyed the conference. It was unlike any neuroscience conference I had participated in—you could feel the sense of interconnection and joy in everyone present, including the organizers, and I felt so nourished by it! Amazingly, though I thought I was there just to give a presentation, I was invited to be a keynote speaker the next year, as well as being invited to be their new Research Director, a position I still hold.

In that first year, 2016, I also applied to give a preconference workshop at the *Toward a Science of Consciousness* conference in Tucson, and while there met with Gary Schwartz, as we had both been invited to speak at a conference on spirituality in Sedona. Gary spontaneously invited me to visit his lab and to give a lecture the next day to his undergraduate class at the University of Arizona. I enjoyed our discussions immensely.

We kept in touch, and about 6 months later, over lunch with our spouses, Gary invited me to co-create with him a new Academy—The Academy for the Advancement of Post-Materialist Sciences (AAPS). When I heard the title of the Academy, I broke into a smile because it seemed so whimsically right—to be part of a new society focused on a post-materialist world order. We agreed that he would be the president, and I would be the vice-president. He also invited me to co-organize with him a first meeting in Tucson of the board of directors, which resulted in the publication of the first book of the AAPS Press titled *The Primacy of Consciousness,* which I co-edited. In all of the situations, I was suddenly interacting with scientists whose research I had admired for years, but never had interacted with before. Each situation seemed like a synchronicity.

About a year later, Gary decided to step down from the presidency, and asked me to take over. I reluctantly agreed, as I have never considered myself an administrative leader. And within a month Gary mentioned that David Lorimer of the Galileo Commission and the Scientific and Medical Network would like to talk about possible cooperative activities between our two organizations, and that in my new role as president, I could move forward with this.

David and I had a wonderful zoom call, and immediately began planning activities together—including two Galileo Commission Summits related to the first two AAPS Press books we had published. I was invited to be a speaker at the *Beyond the Brain* conference, and there met Chris Roe, who synchronistically invited me to contribute to his collaborative project with others, on after-death communication. I took so much delight in these collaborative activities that I was honored when David asked if I would like to join the Galileo Commission Steering Group, and then become co-director of the Galileo Commission. As a result, I am now happy to be able to collaborate with colleagues all over Europe, New Zealand and North America on projects related to educating the public on research about the nature of consciousness.

These amazing opportunities to collaborate with others have continued, with my book *Infinite Awareness* being an impetus for my being chosen to be the Oen Fellow at Luther College in Iowa, where I was asked to talk to students in classes in four departments about consciousness and give a public lecture on my book. The person inviting me told me that he had received an advertising pamphlet from Rowman and Littlefield, my publisher, and before putting it in the wastebasket, thumbed through it, saw Infinite Awareness, was struck by its description, ordered it—and proposed me as a fellow at the College.

And amazingly a composer, Evan Premo, from Vermont recently wrote a choral and instrumental piece based on *Infinite Awareness,* called *The Primacy of Consciousness,* and a portion of it was just performed in early June. It was a powerful experience to be part of that project.

Why do I consider this long series of events to be synchronicities? It is because I felt that I didn't attempt to organize any of the interactions—to make anything happen. I simply wrote that first book, and then joyfully shared my insights and research with others at a few initial conferences. As events unfolded, it seems as if the universe brought me together with person after person, with whom I could collaborate in an ever-widening network of persons all with the same goal—of sharing this research and understanding of the fundamental nature of consciousness with the world. It feels like an ever-expanding circle of love that I have been blessed to be a part of.

Personal significance: This series of events, which has continued to unfold since 2016, leaves me in awe at how a sincere desire of the heart is answered again and again—without the egoic will being involved. There is a desire, then a response from the universe, and this creates joy, and more responses, which seem unrelated to the usual intellectual understanding of cause and effect being linear.

What kind of universe makes synchronicity possible?

I interpret these events according to a model of the universe in which consciousness is fundamental and primary, with the following premises (adapted from papers by Stephan Schwartz (2010) and Bernardo Kastrup (2018)).

- Consciousness is causal, and physical reality is its manifestation.
- All consciousness entities are linked (entangled).
- Consciousness exists throughout the body (and the universe)—so it is even in the physiological processes that appear to affect the mind.
- Aspects of consciousness are not limited by the space/time continuum, and are especially evident when the dissociative boundary of the mind-body complex becomes porous.

Thus, in my first example, I propose that my meditation teacher accessed an aspect of consciousness beyond the space/time continuum, and thus asked me a specific question about getting married. This meant nothing to me until a few weeks later, when I returned to Eugene, Oregon, and met my future husband, with my having a clear sense in that first moment of meeting, that I would marry him. I believe that my own "inner knowing" was also an example of accessing consciousness beyond the space/time continuum, in that this thought has never come to my mind before or since.

I believe the second example is an illustration of the premise that all conscious entities are linked or entangled. Thus, my heart-felt wish to see my hiking friend, allowed a re-arrangement of our activities/schedules, so that we met the next morning on the road near my house, as we were both taking our morning "hikes"—he, on his way to the national forest trails, and me on my usual walk down the local roads. I believe that a key aspect of this example is that the wish arises from the heart, the source of our being, rather than the intellect/egoic framework. I think it is this aspect of ourselves that is truly interconnected with the whole.

The third example relates to many events over a six-year period, which resemble in totality what it must be like to "be in the flow"—that is, making a heartfelt intention, and then welcoming joyfully and responding from the heart to each event in life that comes, as a universal response to that intention. In each of these synchronicities it felt like the "dissociative boundary of the mind-body complex" (Kastrup, 2018) became more porous, allowing a universal creative force to orchestrate events, so that events flowed optimally. And this allowed the manifestation of joyful and often unanticipated outcomes, that were greater than the individual could imagine.

My own philosophical leanings on the nature of the universe are toward what is called idealism. Idealism makes the assumption that consciousness is the universe's foundation, with all natural phenomena reducible to consciousness. Thus, all living beings (including plants and animals) have a conscious inner life, and in addition, there is a consciousness beyond all these individual minds, which is universal. I see the ultimate nature of Consciousness as vast and limitless; it's a vibrant loving awareness, that exists beyond time and space. It is simultaneously the essence of the universe and the essence of our own awareness.

This view includes the following assumptions:

1. Consciousness is causal, and physical reality is its manifestation.
2. All consciousness entities are linked (entangled).
3. Aspects of consciousness are not limited by the space/time continuum.

As this infinite consciousness is the essence and foundation of the universe, I believe it is responsible for bringing this manifest universe into existence, through its own infinite freedom and creativity. And this is part of its way of getting to know and experience all aspects of itself. An ancient text from Kashmir, from the 10th C., the *Pratyabhijña Hrdayam* (Heart of Recognition) says that this act of creation is a play; the term is *kridā* in Sanskrit, and it means that manifestation is a playful, joyful, loving act of universal consciousness, as a way of knowing itself (Wallis, 2017). And, of course, our source and essence *is* this consciousness.

As I believe the essence of each being is that infinite eternal consciousness, it affects my understanding of my own temporal existence and that of other beings in the world. I do not see death as an end of individual existence, but simply a transition to another form of consciousness, with one's core subjectivity untouched. What would survive is the core sense of "I-ness," the one eye that looks out from all creatures. And it could also include the experiences that individuals underwent in embodied form.

Conclusion

As I believe that this infinite consciousness, of which we are a part, is infinitely creative, joyful and playful, and has created the manifest universe as a way of knowing itself (ourself), I see synchronicity as an example of that play. And as all beings in the universe are interconnected, these playful creative acts of knowing ourselves can come in an infinite number of ways.

References

Kastrup, B. (2018). The Universe in Consciousness. *Journal of Consciousness Studies,* 25, No. 5–6, 125–55.

Schwartz, S. A. (2010). Nonlocality and exceptional experiences: a study of genius, religious epiphany, and the psychic. *Explore (NY).* 6(4), 227–236. Jul-Aug. https://doi.org/10.1016/j.explore.2010.04.008.

Wallis, C. (2017). *The Recognition Sutras: A complete translation and explanation of the 1,000- year-old spiritual masterpiece the Pratyabhijñā-hrdaya by Rājanaka Ksemarāja.* Mattamayūra Press.

Woollacott, M. (2015). *Infinite Awareness: The Awakening of a Scientific Mind.* Rowman and Littlefield.

Chapter Eight

Synchronicities: Touch-Points of an Interconnected Reality

JEFFREY DUNNE, PHD

Jeffrey Dunne, PhD is President of the International Consciousness Research Laboratories (ICRL, www.icrl.org), an organization established to continue research on the nature of consciousness carried out at Princeton University from 1979–2007. Beyond his own explorations into the nature of consciousness through ICRL, and several decades of research at Johns Hopkins in fields ranging from acoustics to data science and AI, Dr. Dunne is an award-winning playwright and author. His recently published novel, *Nexus*, weaves the concepts of syntropy and the nature of consciousness into a story that addresses the potentially-existential challenges humanity is facing, highlighting a path by which we can navigate towards a healthier, sustainable future.

Princeton, New Jersey is a town of Old Money. Beautifully manicured, insular, and lush with expensive opportunities, it is the kind of place for people not at all like me. I grew up there, but when my mother passed away, the idea of retaining a house in the area—despite its representation of childhood memories—did not appeal. Yet the place was more than just a childhood home. It was a component of the Princeton Engineering Anomalies Research (PEAR) laboratory ecosystem; it was a piece of the history of ICRL. Turning the asset into a number in a bank account did not feel right.

My wife and I decided that we would retain the *idea* of what the house represented but in a new location. It had to be someplace beautiful, someplace connected to nature where wildlife was welcome and mindless consumerism was not. It needed the right community. After months of research, we felt that the general area around Asheville, NC would be a good fit and set about to find a house there. And we did. It was a beautiful house—secluded yet majestic, on the edge of fairytale in design, and big enough to serve as an ICRL retreat center. We submitted an offer. We negotiated, found an agreeable price, and signed... Well, almost.

At the very last moment, while showing satellite pictures of the place to friends, I noticed a strange strip of pavement about a mile away from the house. Straight, just like... a runway. How had that not come up? People are required to disclose significant things like that, aren't they? We contacted the owners and asked if they ever heard the planes. 'Never,' they replied. We contacted the airport and asked about it. We were informed that pilot training flights, which occur in the summer and autumn as often as 25–30 times a day, have take-off and landing approaches right overhead, and at that point they are at most a thousand feet up. "There's no way they don't hear those planes," the airport official informed us.

The offer never got signed.

We continued our search, but decided to expand. Following the mountain ridges north towards Roanoke, VA, we found places that were even better fits for what we wanted. We selected about twenty to go see and scheduled a trip to visit them. The plans were set. Except they weren't. At the last moment, an event the weekend before the scheduled trip was canceled, and I found myself with a burning desire to get started on the house hunt. Twenty houses, even over a long weekend is not practical. Why not go down the week before as well? So we chose a set that were not too far separated and sent them off to our real estate agent. We couldn't get appointments for all those we picked, so we added one more.

When we arrived at that 'one more', we immediately knew that this was the place. Unlike the other houses, which were on the sides of mountains and had beautiful views in *most* directions, this one was seated in a valley and offered exquisite views in *every* direction. It was modest enough to be easily furnished, yet included an out-building that could be converted into the ICRL center we wanted and many acres of woods for hiking.

While touring the property, our agent received a call from the sellers' agent. The property had another offer, and the owners were in a rush to sell. If we wanted to put in a bid, it had to happen that day. It did. Our offer was not as high as the other, but having sold the Princeton house it was a simple, straightforward cash offer, and that was what the seller needed.

During the process of this short turn-around sale there were a number of other oddities that made the journey... unique. Strange obstructions continued to arise, and as they did, equally strange solutions manifested. A few blurred weeks later the purchase was complete, and we were the owners of a beautiful piece of land in the Blue Ridge Mountains. We were ready to create the new home for ICRL.

About six months after closing, a conference at which I was scheduled to be a keynote speaker (as well as a coordinator) informed me that they had to change everything around. I'd still be giving a talk, but in a parallel session, and there was no reason to show up early because they no longer needed assistance before the conference. Yet my flights to England were already purchased, and changing them would cost far more than a hotel for the corresponding days.

A day after learning all of this I had a phone call with someone I had met through an online workshop two weeks prior. I related the story, and we concluded that this was clearly a sign for doing something else in England for those days. She and I got together, continued our seedling conversations about consciousness, art, and music, and generally enjoyed what London has to offer. While doing so, at one point she casually mentioned her experiences with a group called the Monroe Institute. I recalled hearing about them from decades earlier, but knew essentially nothing about them. A quick internet search revealed something quite curious: the Monroe Institute is located in Virginia. And not just anywhere in Virginia, but right along the drive from my primary home to the new property. Days later, now at the conference, I met someone planning a trip to visit a research group called the Department of Perceptual Studies at the University of Virginia, which, as 'luck' would have it, was also on the drive between my two properties.

Without having any way of realizing at the time, we had managed to select a location for the new ICRL center that was perfectly situated to collaborate with two of the most significant centers for consciousness studies situated on the eastern seaboard (technically three, as I discovered that there was another center only a couple of hours to the south). Quite the... 'coincidence'.

Examining a set of coincidences of this nature, the large number of seemingly random occurrences that led us to a point where we were ideally situated to undertake the next stage of ICRL's growth (despite our ignorance of the factors that would have made that location the logical choice), it is hard to believe that this was just randomness. There are a lot of mountains in the U.S. Why there? To me, there is little doubt that the forces, 'the causes', that led to us buying the property we did are not constrained to the past.

This is but one example from my life, and over the past years I have spoken to many upon many people who have had not merely similar experiences, but even more profound ones. At some point, it becomes impossible to dismiss such overwhelming data as being fabricated patterns out of randomness. Pursuing the possibility that such synchronicities are not simply an illusory trick of a mind seeking to find meaning in a meaningless universe, it is worth exploring the implications that synchronicities are *not* random, and might actually have the potential to provide insights into the nature of reality.

A critical finding of the PEAR research is that our intentions play an important role in the evolution of physical systems. Whether it is a 'random' cascade of marbles through a matrix of pegs, the behavior of light passing through an interferometer, or the digital 'coin flips' of a Random Event Generator, there is an enormous body of data demonstrating outcomes correlated with an operator's intentions. Of course, as with any scientific investigation, we must be careful not to conflate correlation and causality. Are our intentions the impetus by which the outcome emerges, or did the outcome influence us towards having those intentions? Or perhaps both the intention and the outcome are simply correlates from some other driver. It would be naïve to immediately dismiss any of these hypotheses based on the 'usual' interpretation of causation in time given the evidence at hand. Fortunately, we don't need to. All that matters is that the correlation is there, and far beyond any reasonable bounds of being 'just a coincidence'.

Whether one believes that our intentions influence systems—and it is important to note that it has been conclusively demonstrated that those intentions can be held after the systems' behaviors are completed—or that an outcome influences our intentions, we are faced with a contradiction to the traditional view of a unidirectional flow of energy or information in time. There is no avoiding the conclusion that time is far more likely to be a cognitive tool, an organizing principle for the sake of establishing a 'logical' narrative in accordance with cultural expectations, than a fundamentally-invariant prison of an objective universe. The evidence simply doesn't support the materialist perspective.

This conclusion arises from the equations of physics as well, which support the notion of time flowing in both directions. In fact, the same equations that give rise to the principle of Entropy—which predicts that systems will progress from ordered states into disordered states—have another solution that gives rise to Syntropy. Syntropy paints a universe in which information and energy flow backward in time, and in which systems evolve in accordance with achieving some future end state, i.e. a future *purpose*. This is helpful if one is trying to explain how the comparatively simple life forms of amoebas managed to evolve into the considerably more complicated life form of the modern mainstream scientist rather than the other way around.

What seems the inescapable conclusion in all of this is that humans (along with everything else) are part of the much larger, interconnected ecosystem that is the universe. Whether one uses terms like quantum entanglement, karma, or William James' "aboriginal sensible muchness", the message is the same. Far from living isolated lives, we are intrinsically connected to everything else. And as such, it is only reasonable that the touchpoints, i.e. the instances of connection between ourselves and everything else, will have a profound impact on how our lives unfold.

I hypothesize that what we experience as synchronicities—those unexpected coincidences that somehow result in profound life changes and line up the future in such precise and impactful ways—are these touchpoints of interconnectedness. Why did I decide that we should go to Roanoke a week earlier than planned? For that matter, why did our plans for that prior weekend suddenly fall through? The idea that all of that was simply random is only sensible from a perspective that insists that the future cannot influence the past, but repeated scientific experimentation has demonstrated that such is not the case. We've shown that the future *does* have an influence, and consequently the most logical, "Occam's Razor" conclusion is that these synchronicities have significance.

There are several important take-aways from this line of reasoning.

First, recognizing that 'synchronicity' is not synonymous with 'coincidence', we must appreciate that it is impossible to know in any given moment what events may prove to be truly impactful. Consequently, it behooves us to pay more attention to what's going on in our lives, and to exercise restraint in levying judgements about what is good versus bad. It may very well be that something unpleasant in the moment is steering us toward something important... if not toward something wonderful, perhaps toward something necessary.

Second, the presence of synchronicities can provide hints about what is to come. In my experience, I have found that an increased focus on identifying synchronicities has given me a more refined and finely-tuned ability to recognize them. This is not at all surprising. Every skill we develop as humans becomes better with practice; the more we focus on something, the more fluent we become with it.

Third, and perhaps most importantly, synchronicities should serve as a constant reminder that we are part of an organism greater than ourselves. Just as we would not want the bacteria in our stomach waging war against the bacteria in our intestines to determine which will "own" the whole body, so too should it be with people. Synchronicities should be reminding us, in every moment, that everyone is serving a need in the greater organism of humanity (which itself is part of an even greater organism, and so on). As an interconnected, integrated whole, the best thing we can do is align ourselves in the pursuit of promoting the healthiest ecosystem possible.

Chapter Nine

Synchronicity and Unexpected Messengers

STEPHAN SCHWARTZ

Stephan A. Schwartz is a scientist, futurist, and award-winning author of both fiction and nonfiction. He is currently a Distinguished Associated Scholar of the California Institute for Human Science, Distinguished Consulting Faculty Saybrook University, and a BIAL Foundation Fellow. Schwartz is a columnist for the journal Explore, editor of the daily web publication *Schwartzreport.net* and the weekly *Schwartzreport* podcast, in all of which he covers trends and research affecting the future. For over 50 years, as an experimentalist, he has been studying the nature of consciousness, particularly nonlocal consciousness at both the individual and social levels. He is one of the small group that created modern remote viewing, including several now standard protocols. In addition to his non-fiction books and novels, he is the author of more than 350 technical reports, papers, academic book chapters, prefaces, and introductions.

Studying synchronicities for over a half century has left me with two conclusions. First, that many perhaps even most synchronicities are precognitive nonlocal perceptions. An example from my own experience: I had a dream and part of it involved a woman in a very stylish but unusual red dress. In the dream I knew I knew her but could not identify her. When I awoke I tried, but not very successfully, to analyse the dream to find its deeper meaning then, as I began a very busy day, I forgot about the dream. When I got home my wife said there was a movie at our neighborhood theater she would like to see, and we decided to go after dinner. When we got to the theater there was quite a line. As we stood there talking, slowly moving forward, I noticed a woman ahead of us wearing the red dress I had seen in my dream. She and the man with her were about six people ahead of us. As I stared at her she turned and looked back at me, and clearly recognized me although I did not recognize her.

She said something to the man she was with, walked back to me and introduced herself as Barbara. She said that she had seen me in a television interview I had recently done on the Merv Griffin Show and that she was very interested in remote viewing. She told me she had tried it with a friend and been successful, and wondered if I was looking for new viewers for my experiments. We chatted for a moment as the line kept moving, I gave her my card told her to call me, and she went forward to her partner and into the theater. She did call me the next day; I had her come in and did a session with her which was very successful, and over the next few weeks did several more all of which went well. Then I invited her to participate in an upcoming archaeology project I was planning. She would go on to become a regular part of Mobius' remote viewing team.

Clearly in retrospect I saw this synchronicity as basically a precognitive dream anticipating meeting someone the next day who would become a regular participant in my research.

But I believe there is a second kind of synchronicity, one that may involve precognition but which also has a deeper dynamic. To understand this type one must first understand the continuity of consciousness. The idea that we have an eternal self, what all religions recognize, calling it the soul or the spirit, the élan vital, the animus, higher self, and through the centuries a host of other terms. I call it the eternal self, and the evidence from reincarnation, NDE, and other research suggests that this eternal self episodically manifests and incarnates a personality, and that what comes across incarnations is information.

Once one grasps the concepts of the continuity of consciousness and the eternal self, then what I think the research and my own experience is saying is that we make commitments in the nonlocal domain that are basically agreements between two manifesting personalities before either is incarnate. One shares with another a commitment: "At a particular time in this coming life I will come into your awareness to give you a message that will help you." Not that there will be a relationship, but rather that one will serve as an unexpected messenger for the other in a time of stress when such a message will make an important difference. Let me illustrate this with another of my own experiences.

Many of my best memories are associated with nature. Backpacks, rock climbs, hikes, canoe trips, scuba diving. Things as simple as times spent near a lake or just sitting quietly watching a stream flow. I started going into the wilderness when I was about 11, and diving when I was 15. These were the highpoints of my life except for one period of interruption. The period of that interruption began when, in 1971, just after I had resigned as editor of *Seapower* Magazine to write a book when, to my surprise, I was asked to come into government service and become a Special Assistant to the then Chief of Naval Operations, Admiral Elmo Zumwalt. It was the height of the Viet Nam War, a war I had thankfully avoided because I had served in the army before it began. I thought it a grievous geopolitical mistake and my first reaction was to decline the offer. However, when it as explained to me that I was being asked to join a small team of people who had been tasked with restructuring the American military from an elitist conscription force into one that was all-volunteer and racially and gender equalitarian, I agreed, having no real idea what I was getting into.

Senior staffs in Washington all take what seems to me in retrospect a perverse pride in the martyrdom of their overwork. They know it leads to burn-out, excessive partisanship, divorce, and alienated children; but they do it anyway because they are playing a role in shaping American society and the world. Intellect and ego anesthetize the emotions with the narcotic balm of self-importance. Even as I recognized the process was happening to me, I found I was no better able than my neighbor to resist the siren call, and it became all-consuming. Weekends, week nights, all were sacrificed on that altar. There were no times for walks. Even when I jogged along the Potomac River, I found no pleasure in water, wildfowl, or flowers. My life became almost entirely mental and interior, and I found that the subtle connection I had once felt with nature was being broken.

In the Spring of 1972, Admiral Zumwalt and John Warner, the Secretary of the Navy, for whom I also did some work, were both scheduled to be out of town at the same time, and their absence gave me the opportunity to plan a backpack in the Appalachian Mountains. It had been on the books for weeks when I got a call inviting me to speak at a Quaker conference for young people to be held in Maryland that same weekend. My rationale for accepting was that I had been asked to explain the radical and highly controversial programs we were developing to make the American military an all-volunteer meritocracy blind to race, religion or gender. The invitation appealed to both my ego and my heart. I was, and am, proud of this work, and the man who invited me used that pride to convince me to accept. I scrapped my plans for the backpack, and then faced the realization that once again, however valid the reason, I had sacrificed my contact with nature.

Five of us had been invited as speakers, none of whom I knew except by reputation like the Reverend William Sloane Coffin, chaplain at Yale, a well-known anti-war pacificist. When the conference began there were about a fifteen hundred college and high school students in the audience. We spoke one at a time, each of us trying to make some sense of the Viet Nam war, and the events tearing the country apart. What struck me most was that no matter how hard we tried, no matter how sincere we each were, none of us, neither audience nor speakers, seemed able to move very far from the fixed positions that defined our national struggle. It was an argument that had become for me both overly familiar and sterile, and as I listened to it. I became depressed over the opportunity I had missed to get away to the mountains.

On the last day, I was standing talking with Paul Desjardins, a professor of philosophy from Haverford College, near Philadelphia, when the conference organizer asked us if we would do a panel together, just to answer questions for an hour from the floor. I looked at Desjardins, a small neatly made man with a wonderful laugh, and a well-developed appreciation of the absurd, the only other speaker with whom I had made any real contact. He nodded, and we agreed.

Early in the panel, a young woman came to the microphone in one of the aisles and said she had a question for Desjardins. Her tone and her body language were intense. "Don't you believe that these conferences," with a wave of her hand around the room, "all of this is meaningless? Just a way to keep the masses happy and compliant, while the war machine rolls on. Don't you agree that we should leave the cities and get back to the land?"

Desjardins took time to reply. "No, I don't think I do agree," he said. "I'm a college professor. Colleges are by definition collegial, and that requires a concentration of people. They seem to do better in cities, or at least good sized towns. But..." and here he paused, seemingly gauging his answer as much for himself as the audience, "because I cannot move to the countryside does not mean that I cannot go to the mountain. In fact, I go every morning to a memory of my time in the moutains when I meditate. Living in the city has never stopped me from visiting nature in my mind."

The young woman, clearly dissatisfied but non-plused with this answer, started to ask another question, but others were pushing to get to the microphone, so the exchange moved on. The next question was for me, and I have no idea how I responded, because the implications of Desjardin's comment had already begun to sink in.

Half an hour later, as the question period was coming to an end, a young man made his way to a microphone, and when it came his turn, he directed his question to Desjardins.

"Professor, you said in your talk that you were a Roman Catholic. So am I. That's why I am directing my question to you. Don't you think that the rituals of the Catholic Church are empty and irrelevant? I mean as we look around at the pain and suffering in the world how can these meaningless medieval ceremonies have any relevance to a world such as ours?"

They looked at each other for one long beat, staring across the hall, then Desjardins said "No, I think there is still purpose to those rituals. I undergo the rite of baptism every time I take a shower. I take communion every time I break a piece of bread."

We all left later that afternoon, and I never saw Paul Desjardins again. Looking back, I don't know what others thought about him, or his comments but, from that day, I have used the unexpected message he left me with, and rededicated myself to fostering wellbeing every morning when I shower, and I do not eat anything without pausing for a moment to say a kind of nonreligious grace. And when I am overloaded with work and can't take a break I begin my nightly meditation with my wife by recalling one of my memories of my trips to the mountains. Making small mundane acts like these sacred has become my key to staying in touch with who I am, and my commitment to foster wellbeing in every act. All thanks to Desjardins, an unexpected messenger who came into my life synchronistically at a time of great stress to show me something that has changed my life for the better ever since.

Chapter Ten

Synchronicity: So What?

LAUREL WATERMAN, PHD CANDIDATE

Laurel Waterman is a Ph.D. Candidate in Curriculum and Pedagogy, Wellbeing Emphasis, at the Ontario Institute for Studies in Education (OISE) at the University of Toronto. She teaches narrative non-fiction writing at the University of Toronto and is an adjunct faculty member in the Alef Trust's MSc in Consciousness, Spirituality, and Transpersonal Psychology. She is a board member of The Scientific and Medical Network (SMN) and a student member of the Academy for the Advancement of Postmaterialist Sciences (AAPS Global). Laurel's research focuses on consciousness education, aiming to popularize participatory and postmaterialist paradigms of consciousness in education.

I did not used to believe in synchronicity. I thought that corresponding events were only chance coincidences that ultimately mean nothing beyond the meaning that people ascribe to them. For if time moves in a line from the past to the future, as Newtonian physics assumes, and if life on this planet exists because of an accidental collision of elementary particles, as implied by The Big Bang, and if "existence precedes essence," as Sartre proclaimed (1946, in Aho, 2023), then there could be no mechanism by which distinct coinciding events could have a meaningful relationship. I grew up enculturated by scientific materialism, which believes that matter is the primary reality, the brain produces consciousness, and when the body dies, that is the end of one's consciousness. This conditioning runs deep. To this day, I still secretly roll my eyes at most of the stories of synchronicity that people share. I think my stubborn skepticism reveals something important about the nature of synchronicity. I will return to this idea. First, I will share a story from about six months following my husband Aaron's death when I first experienced synchronicity.

The penny story (a version of this story was previously published in Waterman, 2022)

January 2019 in Toronto, Canada. I was deep into my first winter without Aaron. That meant celebrating his 42nd birthday, Hannukah, and New Year's Eve without him, shoveling all the snow myself, and being the only one to tie up the kids' skates and hunt for their matching mittens. Our names finally rose to the top of the waitlist for Dr. Jay's Children's Grief Centre. The intake appointment was on a Thursday evening. It was particularly cold, around -30 degrees Celsius, and I was particularly distraught. Something about walking into a children's grief centre with your six- and four-year-olds who have lost their dad makes the grief stab even sharper and deeper than usual. The kids followed one therapist to an art therapy room, and I sat with another therapist named Heather. I wept throughout the session. I remember saying to Heather, "I just need to know that Aaron is still out there and that he can see us. I feel like if I knew that, I could find the strength to go on."

Heather said, "But nobody knows that. Why can't you just believe it?"

I said, "I wish I could, but I can't. I need to know. I need proof. Not necessarily scientific proof, but personal proof, beyond any doubt." Heather looked at me with sympathetic eyes. The session ended. I wrapped the kids up in coats and scarves, and we navigated the underground parking maze to find the car. Aaron used to do most of the driving. Tears ran down my cheeks as I drove. What happened next, I wrote about in my journal:

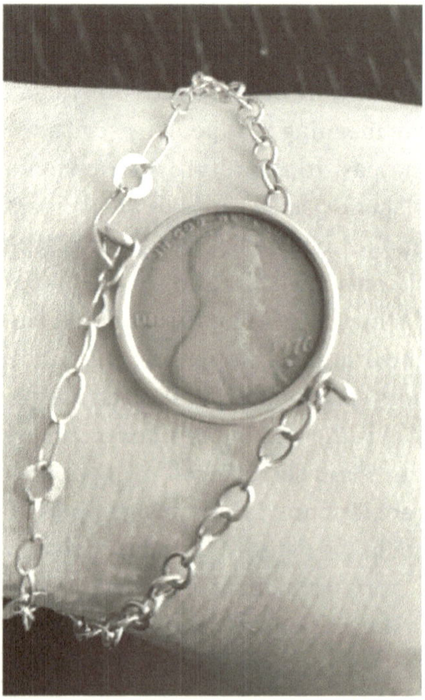

After the grief session at Dr. Jay's, Althea [our four-year-old] *fell asleep in the car on the drive home. I let Izzy into the house and walked back to the car to carry Althea up to bed—Aaron's job. As I approached the car door, I heard a coin drop on the ground. A thought entered my mind:* What if that coin is from Aaron's birth year? Wouldn't that be proof? *I looked down*

and saw a penny in the snow. I picked it up. An American penny. I squinted in the darkness to read the date. I couldn't believe it. I squinted harder and lifted the penny into the streetlamp's light. 1976. Aaron's birth year. And a thought again: Here's your proof. I'm fine. Be happy! (Journal entry, January 31, 2019).

I squeezed that penny and my heart lifted. I smiled for what felt like the first time in months. I carried Althea into the house and placed the penny on the mantel. Later, I had the penny set in a bracelet. I always wear it to remind myself of that powerful sign.

Finding meaning

Jung coined the term synchronicity to describe the experience of "meaningful coincidence" (1952, p. 827) in which an inner psychic state relates to an outer physical event not causally but through the meaning that the events jointly express; Jung also used the term to describe the acausal connecting principle that explains why these kinds of experiences occur (Main, 2019, p. 9). In other words, experiences of synchronicity illustrate the principle of synchronicity. If one accepts that something beyond random chance may be involved in synchronicity, the phenomenon may reveal something about the nature of reality and the relationship between matter and psyche.

I had not considered these ideas at the time of the penny story, yet the number of variables that coincided seemed meaningful. The timing was one thing, occurring not an hour after I cried aloud to a therapist that I just needed proof that Aaron still existed. The rarity of finding a penny from 1976 was another, especially since pennies have been discontinued in Canada and rarely circulate. We had been in LA with my brother over Christmas that year, which accounts for having an American penny. Did it fall from my coat pocket? And how could I hear it fall in the snow? Nevertheless, what gets me is the thought that entered my mind when I heard the coin fall. Ordinarily, in that situation—on a freezing, dark weeknight past bedtime, trying to get my kids to bed—I would not stop to look for a coin in the snow. The thought compelled me to stop and look. This was the first time I had ever considered the date on a coin. (I admit that now I periodically inspect coins for their dates in search of meaning. As an aside, I have not found a coin from 1976 since). That the date on the coin confirmed my thought about Aaron's birth year stretched the boundaries of coincidence. This evidenced a connection between an inner psychic state (a thought) and an outer physical event (the penny). But what pushed it over the edge was the feeling I experienced from this event: a communication from Aaron ("Here's your proof! I'm fine. Be happy!"), and a wash of love and comfort so powerful that a smile broke through the depths of my grief. The emotion is part of the meaning.

That said, this experience alone did not convince me of anything. Other significant events occurred—an uncanny detail from a medium, an experience during an ayahuasca ceremony, and an experience with psychometry. It was the combination of several experiences and piecing them together into a narrative that eventually led to a shift in my ontology (Waterman, 2022). Based on what I had learned through my experiences, stories of others' experiences, and consciousness studies research literature, I could no longer rationally ascribe to a materialist paradigm. In phenomenological and empirical research data, there are just too many "anomalies"—occurrences that materialist theories of consciousness cannot explain (Powell, 2020, p. 281). My favorites include veridical out-of-body experiences during NDEs (van Lommel, 2021; Woollacott & Peyton, 2020), *psi* phenomena (particularly remote viewing (Schwartz, 2020) and precognition (Sheehan, 2023)), savant syndrome (Powell, 2020), terminal lucidity (Roehrs et al., 2023), and young children's memories of past lives (Mishlove, 2021). Quantum field theory, which dismantles materialism's assumptions of reductionism, determinism, mechanization, scalability, and absolute time and space is also mindboggling (Barušs & Mossbridge, 2017). Any one of these phenomena that I have experienced or read research literature about alone might have prompted my curiosity; taken together, they have pushed me into the ontological territory of postmaterialism. According to this paradigm, "mind is fundamental to the universe, i.e., it cannot be derived from matter and cannot be reduced to anything more basic," and "there is a deep interconnectedness between mind and the physical world" (Beauregard et al., 2014, p. 273).

An appealing aspect of postmaterialism is that it leaves space for the mystery. Postmaterialists sense there is something more to reality than the material and admit to not understanding exactly what that something is or how it works. Realizing that there is something more has been profound enough for me. My prior assumptions about the relationship between mind and matter, space and time, life and death, and the meaning of life all shattered. I do not have answers, but the questions alone fill life with new possibilities.

With this ontological shift that allowed for synchronicity, I began to invite synchronicities. I bought the book *Signs* (Jackson, 2019) for my step-daughter, Hannah, and I quickly read it before giving it to her. The author recommended asking deceased loved ones for specific signs. One morning, as I was walking home from dropping the kids off at school, I wondered what sign I should ask for. I thought of Aaron's company, WineOnline.ca, now run by his brothers, and asked for the delivery van to drive by me. After I had walked another block, I realized that I would be at home for the day without much chance of seeing a van drive by, so I changed the sign to the Wine Online logo. I arrived home and spent the morning working. After lunch, I decided to treat myself to the last piece of Linzer torte. This was a special pie usually cooked by Aaron's mother, but she lived far away. Tina, my sister-in-law made this one. I took the last wedge of the pie and crumpled up the parchment paper. The Wine Online logo stared back at me. Tina had cut a circle around a Wine Online delivery box logo and used it as a pie plate. A feeling of joy hit me. Here was my sign, just a few hours after I had asked for it, in our home, under the last piece of Aaron's favorite pie.

Another time, I was preparing to host a big Rosh Hashana dinner with Aaron's side of the family. Rosh Hashana was important to Aaron, and I missed him a lot as I waited in the long line at the Jewish bakery to buy the traditional round challah. When I left the store, I cried for a while, then walked my dog by Aaron's memorial tree. I stopped at the tree and asked Aaron to please show up for dinner—somehow, in some way. As I walked on, I considered buying flowers for the table. Aaron loved flowers and would spend lavishly on them, but I had spent enough money on the food that I decided not to indulge in flowers. My next errand was to go to the fruit market to buy apples to go with the honey—a symbolic food signifying a sweet new year. I only said hello to the clerk as I paid for the apples, and she handed me a bouquet. She insisted I take the flowers with no charge. I felt Aaron smiling. Then later, Aaron's teenage daughter, Hannah, who had said she could not come to the dinner, surprised us by showing up.

If someone else told me these stories, I might silently dismiss them as coincidences. However, because I experienced them, I understand the rarity, the emotional context, the timing, and the layers of thought and matter that must have conspired to orchestrate these meaningful moments. These three stories of synchronicities that I shared here each occurred within hours of a yearning to confirm the continued presence of a deceased loved one. What kind of universe makes these types of synchronicities possible?

Implications

Butzer suggests that synchronicity may offer evidence to support an ontology in which consciousness is primary to matter (Butzer, 2021). Based on my experiences, I wonder if synchronicity also provides clues about the nature of this primary consciousness. The circumstances of the synchronicities I shared suggest that consciousness has a benign quality. Perhaps we live in a universe that cares. For it was at times when I needed comfort that these moments of apparent connection between inner and outer reality supported me in developing the awareness that there is more to life than the material.

Now, one might ask, and in fact, my doctoral supervisor did ask, "So what?"

So what? My prior assumptions about reality no longer make sense. Schools reproduce an ontology that cannot explain the reality we experience and observe. Young people are taught that the brain produces consciousness, which implies that there is no continuity of consciousness beyond this one lifetime, no soul, and no reality beyond this material world when the data suggests otherwise. Imagine if, instead, young people learned about the mystery.

This is the thrust of my research in consciousness education, which explores perspectives on the source and nature of consciousness and their implications for ways of being, knowing, teaching, and learning. If we consider synchronicity from a postmaterialist perspective and grant that something seems to be happening beyond individuals making meaning out of random coincidences, then this challenges several epistemological assumptions that underly mainstream education. Examples of such assumptions include: 1) thoughts alone do not influence material reality; 2) thoughts cannot be communicated without the five ordinary senses; 3) the brain produces mind—the source of intelligence, creativity, problem-solving, and emotion; and 4) Anything without a functioning brain does not have consciousness. However, postmaterial interpretations of synchronicity imply a bidirectional relationship between thought and matter, and an interconnectedness between inner and outer worlds (Butzer, 2021). This flips the above assumptions. Somehow, it seems that perhaps 1) thoughts can or do influence material reality; 2) thoughts may be communicated without the use of the five ordinary senses; 3) mind exists, perhaps even originates from beyond the brain and body; and 4) consciousness can exist or survive without a functioning brain. What might teaching and learning look like if we center these assumptions?

In "Standing in the Mystery: Holism as Interconnected Consciousness," (Waterman & Nigh, forthcoming), Kelli and I write about synchronicities that occurred in our classrooms and how, when integrated, they point us to different ways of being as teachers. We invite in the qualities of the witness who does not know, a trust in the whole that supports us, and the awareness of the invisible world and our connection to it. These spacious approaches to teaching are supported by initial data from my doctoral inquiry in process, a narrative study with scholars about their experiences teaching about consciousness beyond the brain in university settings. Several participants have emphasized the importance of leaving room in their lectures, lesson plans, and course designs for synchronicity. This pedagogical approach differs from the outcomes-oriented, individualistic, banking concept of education (Freire, 2011) that is still prevalent in schooling at all levels.

In "Consciousness Education: Why an enquiry into consciousness is an educational and spiritual imperative" (Walton & Waterman, forthcoming), Joan and I outline ideas for curricular approaches to consciousness education. These include inviting stories of "anomalous" or "spiritual" experiences into the classroom through reflective, contemplative, and creative practice; working with attention and intention as a potential source of healing, inspiration, intuition, and creativity; engaging with learning fields (Bache, 2008); and exploring interconnected relationality, a concept from Indigenous scholar Aileen Moreton-Robinson "where one experiences the self as part of others and that others are part of the self. This is learnt through reciprocity, obligation, shared experiences, co-existence, co-operation, and social memory" (Moreton-Robinson, 2013, p. 341). This relationality extends to the natural world, which, with these shifted assumptions, we understand to be brimming with consciousness.

Conclusion

In this chapter, I shared personal stories of synchronicities that, taken together with other experiences and research literature, have convinced me of the principle of synchronicity: that there is some mysterious, acausal connection between psyche and matter that centers around meaning. A challenge in communicating about synchronicities is that the meaning piece often gets lost in translation. My stories of experience, no matter how well I write them, will fail to convey to you the meaning of these experiences for me. This is part of why I am still so stubbornly skeptical when I hear of others' stories of synchronicity; I cannot fathom the meaning for the experiencer. And I suspect that the meaning is the point. When I moved from the stories to their implications, I took on the question, *so what?* and relayed how the shift in my ontology affects my epistemological assumptions, and how these play into ideas about consciousness education. Then I cut out a paragraph about how these shifts in assumptions and approaches might help us to solve significant problems in the world, like the climate crisis, the wellbeing crisis, and social and ecological injustice. I do believe that an ontological shift from materialism to postmaterialism is essential if we are to solve these wicked problems. However, I cut it because it goes beyond the point. Synchronicity: *So what?* Because it proves the existence of meaning.

References

Aho, K. (2023). Existentialism. In E. Zalta & U. Nodelman (Eds.), *The Stanford Encyclopedia of Philosophy* (Summer Edition). https://plato.stanford.edu/archives/sum2023/entries/existentialism/

Bache, C. (2008). *The living classroom: Teaching and the collective unconscious.* SUNY Press.

Barušs, I., & Mossbridge, J. (2017). T*ranscendent mind: Rethinking the science of consciousness.* American Psychological Association.

Beauregard, M., Schwartz, G. E., Miller, L., Dossey, L., Moreira-Almeida, A., Schlitz, M., et al. (2014). Manifesto for a post-materialist science. *Explore, 10*(5), 272–274. doi:10.1016/j.explore.2014.06.00

Butzer, B. (2021). Does synchronicity point us towards the fundamental nature of consciousness?: An exploration of psychology, ontology, and research prospects. *Journal of Consciousness Studies,* 28(3–4), 29–54.

Freire, P. (2011). The banking concept of education. In E.B. Hilty (Ed.), *Thinking About Schools* (pp. 117–127). Routledge. https://doi.org/10.4324/9780429495670-11

Jackson, L.L. (2019). *Signs: The secret language of the universe.* Spiegel & Grau.

Jung, C. G. (1952) Synchronicity: an acausal connecting principle. In *Collected Works,* vol. 8, *The Structure and Dynamics of the Psyche,* 2d ed., (1969). Routledge & Kegan Paul.

Main, R. (2019). Synchronicity and holism. *Revue de Psychologie Analytique.*

Mishlove, J. (2021). Beyond the brain: The survival of human consciousness after permanent bodily death. *Bigelow Institute for Consciousness Studies Essay Competition.* https://www.bigelowinstitute.org/docs/1st.pdf

Moreton-Robinson, A. (2013). Towards an Australian Indigenous women's standpoint theory: A methodological tool. *Australian Feminist Studies, 28*(78), 331–347.

Powell, D. H. (2020). Beyond materialism and madness: A neuropsychiatrist's perspective on anomalous experiences. In S.A. Schwartz, M. Woollacott, and G.A. Schwartz (Eds.), *Is consciousness primary?* (pp. 280–336). AAPS Press.

Roehrs, P., Fenwick, P., Greyson, B., Kellehear, A., Kothe, K., Nahm, M., Roe, C., Tassell-Matamua, N., Woollacott, M. (2024). Terminal lucidity in a pediatric oncology clinic. *Journal of Nervous and Mental Disease, 212*(1), 57–60. doi: 10.1097/NMD.0000000000001711.

Schwartz, S. A. (2020). Mind rover: Exploration with nonlocal consciousness. In S. A. Schwartz, M. Woollacott, and G. E. Schwartz (Ed.), *Is consciousness primary?* (pp. 233-280). AAPS Press.

Sheehan, D. (2024). Thinking backwards about time. *Timing & Time Perception,* 1–5. 10.1163/22134468-bja10103.

van Lommel, P. (2021). The continuity of consciousness: A concept based on scientific research on near-death experiences during cardiac arrest." *Bigelow Institute for Consciousness Studies Essay Competition.* https://www.bigelowinstitute.org/contest_winners3.php

Walton, J. & Waterman, L. (forthcoming). Consciousness education: Why an enquiry into consciousness is an educational imperative. *Journal for the Study of Spirituality.* Taylor Francis Online. (under review)

Waterman, L. (2022). Stories of a transformation in consciousness: A self-study to ground narrative inquiry research in consciousness education. *International Journal of Transformative Research, Special Issue: The Implications of a Postmaterialist Worldview for Education and the Humanities, 9,* 27–39.

Waterman, L. & Nigh, K. (forthcoming) Standing in the mystery: Holism as interconnected consciousness. In K. Nigh and D. Can, *Holistic Education: Enduring principles and practices,* book two. Information Age Publishing (in press).

Woollacott, M., & Peyton, B. (2021). Verified account of near-death experience in a physician who survived cardiac arrest. *Explore, 17*(3), 213–219. doi:10.1016/j.explore.2020.03.005

Chapter Eleven

Synchronicities and the One Mind: The "Theopsychobiophysics" Synchronicity

GARY E. SCHWARTZ, PHD

Gary E. Schwartz, PhD is Professor Emeritus of Psychology at the University of Arizona where he served as Professor of Psychology, Medicine, Neurology, Psychiatry, and Medicine, and as Director of the Laboratory for Advances in Consciousness and Health. He earned his PhD at Harvard University, and previously served as a tenured Professor of Psychology and Psychiatry at Yale University, Director of the Yale Psychophysiology Center, and Co-Director of Yale Behavioral Medicine Clinic. He has published more than 500 scientific papers and chapters, including six in the journal Science, co-edited more than dozen academic books and authored more than a dozen books for the general public, including *Extraordinary Claims Require Extraordinary Evidence: The Science and Ethics of Truth Seeking and Truth Abuse.* He co-founded the Academy for the Advancement of Post Material Sciences and served as its first President.

Introduction

In my 2017 book, *Super Synchronicity: Where Science and Spirit Meet,* I focused on the phenomenon of 'serial coincidences'—what I termed Type III synchronicities or 'super synchronicities[1]'—where six or more highly connected events occur within a relatively short period of time (e.g. a few days). Because of the highly improbable nature of patterns of six or more serial events occurring by chance (systematically calculated using conditional probability

1 I sometimes refer to super synchronicities as 'supersynchroniticies'—one word—similar to concepts like supercomputers, superconductivity, superimposition, supercollider, etc.

statistics), Type III synchronicities provide compelling evidence for complex interconnectedness, higher intelligence, and intentional agency in human (and animal) life.

In an essay of around 3,000 words, it is not possible to adequately analyze the rich, creative, and playful intelligence expressed in such extraordinary patterns[2]. Instead, I will focus here on a more limited set of four recent events (termed Type II synchronicities when three to five highly connected events occur within a relatively short period of time). This particular Type II synchronicity—the 'Theopsychobiophysics' Synchronicity—speaks to the thesis that certain levels of synchronicity may express complex, dynamical, ongoing human relationships with a Universal Conscious Intelligence known historically by many names (e.g. Yahweh, God, Allah, Great Spirit, Source, Divinity, Infinite Intelligence) as illustrated in Dr. Larry Dossey's 2013 book *One Mind: How Our Individual Mind is Part of a Greater Consciousness and Why it Matters.*

The context

To appreciate the personal significance for me of this particular Type II synchronicity, I should briefly explain that (1) I was raised by essentially atheist parents in the 1950's, (2) I was educated in western-based materialist science at Cornell University and Harvard University in the 1960's, and (3) I performed leading edge mainstream research in psychophysiology, health psychology, and behavioral medicine at Harvard University and Yale University in the 1970's and 1980's[3].

However, like many scientists of my generation, my 'faith' in the materialist perspective was slowly but surely being challenged by (1) emerging findings: for example, in physics and astrophysics revealing complex orders and organization in all levels of the universe, from subatomic particles and quantum systems to super clusters of galaxies and the cosmic web, plus (2) direct personal experiences; in my case, observations of complex sequences of precognitive dreams by Christopher Robinson, known in England as the 'dream

2 It is not possible in a short essay to review the history of publications—both for scholars and for general readers—that consider the relationship between complex synchronicities and the existence of a Universal Consciousness Mind.

3 Professor of Psychology and Psychiatry, Director of the Yale Psychophysiology Center, and Co-director of the Yale Behavioral Medicine Clinic (1976–1988).

detective,' as well as replicated personal experiences of super synchronicities.

I first confessed this controversial evidence in my 2006 book *The G.O.D. Experiments: How Science is Discovering God in Everything, Including Us.* The acronym G.O.D. stands for 'Guiding Organizing Designing' process. However, for many reasons I choose to minimize my research activities addressing the One Mind/G.O.D. Hypothesis and decided to wait until I retired from university research and teaching in May, 2024, before focusing on this question.

Though I have mentioned the One Mind/G.O.D. hypothesis in a number of recent books and papers including the *Super Synchronicity* book, the 'deciding moment' for me to finally 'go public' only occurred in the spring of 2023. The triggering event was an upcoming pair of conferences I was co-hosting with Dr. Thomas Plante and Catherine Wolff titled *Faith Seeking Understanding: Conversations on the Emerging Scientific Foundations of Faith.* These conferences were inspired by the combination of an academic book Dr. Plante and I had co-edited in 2022 titled *Human Interaction with the Divine, the Sacred, and the Deceased: Psychological, Scientific, and Theological Perspectives,* and a book written by Catherine Wolff for the general public in 2021 titled *Beyond: How Humankind Thinks About Heaven.* The conferences were held on May 19, 2023 at Santa Clara University and May 20, 2023 at Stanford University.

The four synchronicity events related to 'Theopsychobiophysics'

It was around 3 am Thursday morning, April 13, 2023 and I was 'reflectively meditating' about the upcoming *Faith Seeking Understanding* conferences when it 'popped into my head' that it might be time for me to 'go public' and to officially announce the birthing of a new interdisciplinary field that I was calling 'theopsychobiophysics' (bridging theology, psychology, biology and physics). It occurred to me that maybe it was time for me to write an overview paper tentatively titled "Bringing the God Hypothesis into Contemporary Science: The Emergence of Theopsychobiophysics." At that moment the organization of this paper appeared fully formed in my mind. This became Event #1.

Startled by this experience, I decided to share these early morning musings with my wife Rhonda as we were having coffee together that morning around 7 am. I wanted to hear Rhonda's thoughts about (1) the wisdom of my helping to create such an interdisciplinary field, and (2) the timing of doing so now. To my surprise, Rhonda was highly supportive of both.

It turned out that the person I wanted to invite to be the co-author of this paper, a person I regularly meet with on Thursday mornings at 9 am who is doing a part-time postdoctoral fellowship with me bridging science, spirituality, and religion (Dr. Michael Lafasaki), had to cancel our meeting that morning. This fortuitous circumstance freed time for me to act upon my early morning musings plus Rhonda's encouragement and write a draft of an abstract as well as an outline of the possible paper. This became Event #2.

After completing the draft of the abstract and outline, I immediately emailed the abstract to Dr. Lafasakis and explained how the combination of unanticipated Events #1 and #2 had made it possible for me to (1) see the vision and make the decision to move forward with theopsychobiophysics, and then (2) formalize the vision into an abstract plus detailed outline. I shared the draft with him and invited him to collaborate on the article.

I should explain that at this point I was not thinking about synchronicities. It was what happened next that triggered my awareness of a potentially emerging serial synchronicity related to theopsychobiophysics, and it literally took my breath away.

It was around 11 am, and I had just emailed a draft of the abstract plus outline to Dr. Lafasakis. To my surprise, Rhonda then walked into my office holding a large envelop addressed to me that she had just retrieved from our mail box. As soon as I saw the envelop, I realized what it might be. A couple of weeks earlier I had ordered a personalized license plate that was going to replace my license plate that said 'SOULSCI' (short for 'Soul Science'). Rhonda and I had three cars at that time, and this license plate was to be placed on our 'vacation/daycation' car. Since I was pondering my future retirement from the university, and I wondering if it was time for me to start focusing on the One Mind/G.O.D. hypothesis, I thought I would remind myself of this goal by changing the plate from 'SOULSCI' to 'GOD SCI'[4]. This was especially the case because I would be giving a presentation at the Santa Clara University conference with a title provided by Dr. Plante—"Review of Communication with the Unseen: Scientific Evidence"—he wanted me to include credible evidence pointing to communication with the Divine.

4 Arizona license plates can take a maximum of seven letters and spaces (some plates only allow six or even 5). Since 'SOULSCI' was seven letters, no space was possible; however, 'GOD SCI' was only six letters, hence the space.

Above is a photo of the 'GOD SCI' plate on my car. The photo was taken on March 27, 2024 for this essay.

It was the exquisite timing of (1) having just finished the draft of the theopsychobiophysics abstract and emailing it to Dr. Lafasakis, and then (2) receiving the 'GOD SCI' plate via Rhonda at 11 am, that left me speechless. I was fully mindful that this coincidental timing could have been just that—a random, chance event. However, the significance and salience of:

(1) my envisioning the theopsychobiophysics paper the night before and making the decision to write and hopefully publish it (Event #1),

(2) the unanticipated cancellation of my meeting with the paper's to be invited co-author, giving me an unexpected two-hour block of free time to formalize an abstract plus outline of the paper (Event #2), immediately followed by

(3) the surprise timing of Rhonda coming into my home office with the GOD SCI plate (Event #3), seemed especially meaningful to me. However, it was Event #4 that happened a couple of days later that solidified my interpretation that this was an especially meaningful Type II serial synchronicity.

In brief, what happened was that upon receiving the GOD SCI plate, I was spontaneously inspired to share the working title and abstract with Dr. Plante, and I asked him if he would be interested in being a third author on the paper. He immediately wrote back (he is typically a very fast email responder) and indicated that he would be delighted to be a collaborator. This set the stage for my discovering Event #4.

As fate would have it, Dr. Plante and I had a Zoom meeting scheduled for the following day to discuss our upcoming conference. The new item I wished to add to our agenda was to ask him if he felt it was appropriate for me to end my overview presentation with a brief introduction to the possible birth of theopsychobiophysics?

Since Dr. Plante did not know much about Dr. Lafasakis, I shared some of Dr. Lafasakis's background, including the fact that besides doing a part-time postdoctoral fellowship with me (which we anticipated would take five years), that Dr. Lafasakis had recently been admitted to Holy Cross University's part-time Greek Orthodox Diaconate Program (which he anticipated would also take around five years). Dr. Plante then essentially said, and I paraphrase, "you're not going to believe this." Dr. Plante revealed that for many years he has served as the outside evaluator for prospective students west of the Mississippi river who apply to the Holy Cross University Diaconate Program (and Dr. Plante is a Jesuit Catholic)!

What are the odds? This highly unlikely, extraordinarily meaningful connection became Event #4.

Personal significance

As I confessed in my 2006 book *The G.O.D. Experiments,* some of the most compelling evidence for intelligence and guidance in the orchestration of the universe can be found in the phenomenon of synchronicity. A synchronicity requires a minimum of two events (a Type I), and they are the most common expression of highly improbable plus highly meaningful coincidences. Here the term coincidence refers to relative co-occurrences, since the 'co-occurrences' are really serial events occurring over time.

My personal synchronicity journey began while I was at Yale University where I discovered 'too many elevens' in my life at the time. In fact, there were more than eleven events involving the number 11. In trying to make sense of that statistical oddity, I ended up learning about early Jewish mysticism, Kabbala and numerology, followed by some astronomically improbable and highly meaningful personal experiences that required that I entertain the existence of some sort of G.O.D. process in the Universe.

I decided to write a chapter describing the synchronicity evidence for the G.O.D. Experiments book. However, my then editor at Simon Schuster, plus my then writing partner, wanted to exclude this chapter because they thought "no one will believe it" and they reasoned that "I will lose scientific credibility." I explained that it was not my fault that the evidence occurred, and that I could not in good conscience publish a book about a G.O.D. hypothesis without including such real-life evidence.

We agreed on a compromise. I could include the chapter titled "Synchronicity in New York City" as an Appendix in the book. At that time, I made myself a promise. I decided that if such super synchronicities continued to point to the presence of what Sri Aurobindo called a "Super Mind" and Dr. Dossey called "One Mind", that eventually I would publish such evidence and include the "Synchronicity in New York City" appendix as one of the first chapters in a super synchronicity book.

Although I continue to observe evidence of Type I, II and III synchronicities in my life, when we invited to pick one exemplar of special personal significance for us, I decided that the most recent Type II synchronicity described above deserved to be featured here.

What kind of universe makes Super Synchronicity possible? A super intelligent, living, and playful one

In an effort to be thorough and fair to possible explanations of a phenomenon (especially complex phenomena), laboratory scientists who are conceptually and theoretically oriented attempt to consider all potentially testable hypotheses. They will not exclude potential hypotheses simply because they are controversial and challenge historical and/or preferred beliefs. I devoted a chapter in *Super Synchronicity* to explaining 12 levels of theories.

However, it turns out that the original abstract I wrote for the theopsychobiophysics article captures my current thinking about synchronicities, especially highly complex serial ones, and it illustrates a possible overarching scientific framework for envisioning high level synchronicity research.

Abstract

The history of science documents the extensive creation and evolution of interdisciplinary fields that bridge two or more domains of fields of knowledge and research. Examples of well-established as well as emerging interdisciplinary fields, listed hierarchically from micro to macro levels, include: biophysics, physical chemistry, biochemistry, psychophysiology, social psychology, political sociology, behavioral economics, social ecology, and exoplanet climatology. Schwartz has participated in the creation and development of multiple interdisciplinary fields, including social psychophysiology, health psychology, and behavioral medicine. The field of behavioral medicine includes multilevel interdisciplinary fields including psychoneuroimmunology and psychoneuroendocrinology. Relatively absent in the evolution of interdisciplinary science is a systematic consideration of potential theological contributions to complex effects documented within and across disciplines (a contemporary exception is the emerging field of neurotheology). In this paper we present the case for bringing the God Hypothesis into contemporary multilevel interdisciplinary research. Using general systems principles, we (1) discuss how it is possible to integrate theories and variables from theology, psychology, biology, and physics, and (2) explain how by adding theological variables to psychobiophysical research, it becomes theoretically possible to account for more of the unexplained variability in findings as well as address seeming anomalies in contemporary science. Examples of future theopsychobiophysical research are outlined. (Schwartz, Lafasakis, and Plante, 2024)

As a co-editor of this volume, I would like to honor one of the contributors, Vasileios Basios, by ending my essay quoting his concluding three paragraphs. They express—more succinctly and beautifully—my personal emerging conclusions:

Consciousness is the fundamental expression of the immaculate power of love, which is the initiator of its creation. Inevitably, given finite knowledge and understanding, there will always be a gap between what is given and definable and what is real as it comes into existence.

The One, multifaceted and multi-aspect reality that contains and is contained within the whole cosmos, the Unus Mundus (ἕν το πᾶν) of the Ouroboros of Jung and Pauli. It holds the interconnected diversity of its parts by being " within" those parts. When harmonious relationships manifest in this dialectic, they reveal the meaning of meaning.

That's why we must always try to express the inexpressible or speak of the unspeakable. To always keep our minds and systems open. For the cosmos embodies the necessary and sufficient conditions for everything to be as it is, including its logic. It is within such an expanded logic that the apparently paradoxical phenomena of synchronicity find their natural habitat. (Basios, 2024).

References

Basios, V. (2024). Harvesting the fruits of the synchronicity field. In M. Woollacott, G.E. Schwartz, and D. Lorimer (Eds.). *The Playful Universe: Synchronicity and the Nature of Consciousness.* AAPS.

Beitman B. (2022). *Meaningful Coincidences: How and why Synchronicity and Serendipity Happen.* Park Street Press.

Dossey, L. (2014). *One Mind.* Hay House.

Plante, T., and Schwartz, G. E. (Eds.) (2022). *Human Interaction with the Divine, the Sacred, and the Deceased: Psychological, Scientific, and Theological Perspectives.* Routledge.

Schwartz, G. E. (2006). *The G. O. D. Experiments: How Science is Discovering God in Everything, Including Us.* Atria Books (Simon & Schuster).

Schwartz, G.E. (2014). God, synchronicity, and postmaterialist psychology I: Proof-of-concept real-life evidence. *Spirituality in Clinical Practice, 1*(2), 153-162.

Schwartz, G.E. (2015a). God, synchronicity, and postmaterialist psychology II: Replication and extension of real-life evidence. *Spirituality in Clinical Practice, 2*(1), 86–95.

Schwartz, G.E. (2015b). God, synchronicity, and postmaterialist psychology III: Additional real-life evidence and the higher power healing hypothesis. *Spirituality in Clinical Practice, 2*(4), 289–302.

Schwartz, G. E. (2017). *Super Synchronicity: Where Science and Spirit Meet.* Param Media.

Schwartz, G. E. (2021). *Extraordinary Claims Require Extraordinary Evidence: The Science and Ethics of Truth Seeking and Truth Abuse.* Waterside Productions.

Schwartz, G. E., Lafasakis, M., and Plante, T. (2024, in preparation). Bringing the God Hypothesis into contemporary science: The emergence of theopsychobiophysics.

Wolfe, C. (2021). *Beyond: How Humankind Thinks About Heaven.* Riverhead Books.

Chapter Twelve

Resting in the Unknown: A Story and (Potential) Explanation of Meaningful Coincidence

BETHANY BUTZER, PHD

Bethany Butzer, PhD writes, teaches, and conducts research in the fields of positive psychology and transpersonal psychology. She received her MA in clinical psychology and her PhD in social psychology from the University of Western Ontario, Canada. From 2013–2015, Bethany was a postdoctoral research fellow at Harvard Medical School, and from 2016–2022 Bethany was a Lecturer in the School of Psychology at the University of New York in Prague. Bethany is currently a Lecturer for the Alef Trust MSc program in Consciousness, Spirituality and Transpersonal Psychology and she is the Assistant Director of the Alef Trust PhD program in Applied Transpersonal Psychology. www.bethanybutzer.com.

The topic of synchronicity has long intrigued philosophers, scientists, and the general public (Hocoy, 2012). Yet debate continues around whether synchronicities are "real" events that have something to tell us about the fundamental nature of reality (Butzer, 2021), or whether they are simply random occurrences that humans misinterpret to provide a sense of meaning in their lives (Hand, 2015). In this essay, I will share three examples of my personal experience with meaningful coincidences, as well as my ideas about the implications of these types of meaningful coincidences for our understanding of consciousness and reality. Throughout the essay, I will use the term synchronicity to refer to meaningful coincidences that follow the criteria outlined by Carl Jung (1973) in his original conceptualization of the term, namely as "the simultaneous occurrence of two meaningful but not causally connected events" (p. 25) involving an internal mental state that occurs within close timing to an external event. I use the term *meaningful coincidence* to refer to a broader conceptualization of synchronicity that includes occurrences such as serendipity (i.e., a fortunate coincidence that occurs due to action that one takes in the world and that may or may not involve an initial mental state that correlates with an external event) and seriality (i.e., a series of events that occur within close proximity to each other but that may not include meaning) (see Beitman, 2022 for a detailed description of various types of meaningful coincidences).

Personal examples of meaningful coincidence

I will begin by sharing three meaningful coincidences that had a tremendous impact on my life. The first took place in 2015, when I decided to attend a personal development retreat that was being led by a teacher who I wasn't overly familiar with. The retreat was called "Soul Fire," and I felt a strong intuitive nudge to attend. But it was expensive, and I was worried that I had made an unwise choice. The day after I paid my first installment for the retreat, I walked out my door and noticed letters spraypainted on the sidewalk in front of my apartment building (see Figure 1).

Figure 1: Soul Fire Spraypaint (the "I'm sorry" text was to the right, but is not pictured here).

The letters hadn't been there the day before, and I was annoyed that someone had vandalized my neighborhood. I took a quick look at the writing, and stopped when I realized that it said, "I will set my soul on fire. I'm sorry." I stood there for a long moment, shocked to have found what seemed to be a direct message telling me that it was ok to attend the upcoming Soul Fire retreat.

The retreat ended up having a profound impact on my life. I met some of my closest and dearest friends there, and I also experienced several meaningful coincidences at the retreat itself. One of these involved meeting four women who had connections in Prague, where I was moving two weeks later. The retreat participants were a small group of 30 women, and the retreat took place in a remote location in the US, so to encounter so many Prague connections was surprising! I also now realize, in hindsight, why the spraypainted message said, "I'm sorry." The soul-led path is not an easy one, and many aspects of my personality and my life burned to the ground in the fire of that retreat. One example is that when I moved to Prague after the retreat, I was forced to let go of many aspects of my identity that I had once held dear. I had left my job at Harvard Medical School, and I had very little paid work. I also left behind friends, family, and my mother tongue in order to live in a new country and culture. The retreat catalyzed a sense of soul sovereignty that required me to stop paying so much attention to others' opinions, and to begin following my inner guidance more strongly—which sometimes resulted in making life decisions that others disapproved of.

Several years later, in 2019, the same teacher announced a new retreat that she was holding, called "Soul Sanctuary." Again, I felt called to attend, but I couldn't afford it (I had $10 in my bank account at the time). I was living in Prague, and the retreat was in the US. I would have had to put all of the flight, lodging, and retreat registration costs on my credit card, and to do so felt like an unwise financial choice. One cold February afternoon, I decided to go for a walk around my neighborhood while holding the intention to receive an answer to my question about whether I should attend the retreat. I called upon my previous meaningful coincidence with spraypaint and said to myself, "If, when walking around my neighborhood today, I see the word 'Soul' in spraypaint, I will attend this retreat even though I can't afford it." I then proceeded to walk along a route that I often take. I paid particular attention to any and all graffiti on the walls and buildings around me. This is something that I do regularly—my husband is a street/graffiti artist, and I often pay attention to the street art around Prague.

As I climbed the hill on a street that I'd walked on many times before, I looked to my right and saw a spraypainted image of a troll with the following words around it: "Give me absolute control on your soul" (see Figure 2). I'd never seen this painting before. Similar to my experience in 2015, I stood there for a moment, dumbfounded. I felt goosebumps rise on my skin and tears form in my eyes. Then I laughed. I registered for the retreat when I got home, and again, it had a deep impact on my life. Even now, years later, I meet every Sunday on zoom with a group of women from the retreat, and we have supported each other through some of the most beautiful and difficult parts of our lives.

Figure 2: Soul Sanctuary Spraypaint

The third meaningful coincidence is one that I shared in an academic article that I wrote about synchronicity (Butzer, 2021). I've decided to reshare it here, as it offers a particularly compelling example of a synchronicity as originally defined by Jung (1973). In May 2019, I wrote about the following dream in my journal:

"I dreamt that I was in someone's apartment and a Rottweiler dog attacked me by biting my arm. I was calling for help, but no one could hear me."

Approximately three months later, my mother was out for a walk on a beautiful day. Out of nowhere, a Rottweiler attacked her from behind by biting her arm. The dog's owner didn't help her, and she ended up walking home by herself, holding her arm in the air, crying. No one helped her along the way. She eventually made it to the hospital, where she had to receive internal and external stitches. She was very traumatized by the experience, and she eventually won a lawsuit against the dog's owner. Out of all of the infinite things that I could dream about, what is the probability that I would have a dream that so closely paralleled what happened to my mother a short time later? An additional point of interest is that I had this particular dream while I was attending the 2019 "Soul Sanctuary" retreat that I described previously.

A few thoughts come to mind when I contemplate the personal significance of these meaningful coincidences. The series of meaningful coincidences that led me to attend the 2015 and 2019 retreats held very real and practical implications that had an impact on my life. Both retreats catapulted me onto a life path that has been profoundly meaningful (and at times difficult), but ultimately fulfilling. If it weren't for my experiences with "synchronistic spraypaint," I probably would have dropped out of the first retreat and never attended the second one. I wouldn't have met the group of women who have become my closest friends and who provide me with a nourishing sense of community. It could also be argued that I might not have had the dream that so closely paralleled my mother's experience with the dog. The personal significance of this dream is perhaps more subtle than my experience with the retreats. The dream didn't help me protect my mother from the dog attack, and in fact, I didn't even remember having the dream until months later, when I happened to be looking through one of my old journals. Why, then, would I have such a potentially "useless" precognitive dream?

I think the answer lies in the nuanced interconnections between all three meaningful coincidences, as well as my journey toward becoming interested in synchronicity as an academic topic. My professional journey with synchronicity has not been an easy one, as I completed my psychology training at a mainstream university, where talk of topics like synchronicity and parapsychology were taboo. In a sense, my professional life was gaslighting my personal life, because the many synchronistic and parapsychological phenomena that I had experienced were dismissed by my colleagues as not being real. In the case of this particular dream, the mental phenomenon so closely matched the physical phenomenon (right down to the breed of dog and the location of the bite) that it is very difficult to shrug it off as a random occurrence. These types of meaningful coincidences give me confidence and motivation to continue pursuing these topics academically, regardless of potential dismissal or judgment by the mainstream. In addition, the fact that I had the dream while attending a retreat that was itself inspired by a meaningful coincidence, gave me further confidence in the idea that when we pay attention and act on meaningful coincidences, it creates a domino effect, or a "synchronistic cascade" in which multiple synchronicities can follow from each other and operate synergistically. In a sense, it is as if we enter a type of "synchronistic flow" that holds the potential for super-synchronicity to occur, as described by Schwartz (2017).

Implications of meaningful coincidence for our understanding of consciousness and reality

My personal experience with meaningful coincidences, as well as my academic explorations into the science of consciousness, lead me to suggest that we live in a participatory universe that is infused with meaning, in which we co-create reality with some sort of force, intelligence, or information that we do not yet fully understand by conventional scientific methods. My personal answer to the question, 'What kind of universe makes synchronicity possible?' is that we live in an intelligent universe, a universe that has infinite creative potential, and that wants to co-create with us, possibly as a way to better know itself. At its core, I believe that this universe is made up of some sort of fundamental whole, where nothing is separate from anything else, and that we experience subject-object distinctions that emerge from this undivided whole because of our unique sense capabilities. In other words, in the same way that a snake can see ultraviolet light whereas I cannot, I perceive mind and matter as separate because I don't have the sensory ability to perceive the fundamental oneness of which we are all part (unless perhaps I engage in deep meditation, ecstatic states, or consume psychedelic substances).

The concept of meaning is important with regard to the ways in which I conceptualize synchronicity in relation to my ontological understanding of reality. My personal and academic explorations of synchronicity have led me to propose that meaning occurs on multiple levels in our universe. On one level, meaning is a cognitive strategy that humans use in order to create an internal narrative that helps us feel that our life makes sense. On another level, if we think of the fundamental basis of reality as being some sort of interconnected whole and/or intelligence, then meaning could be the basis or creative impetus of this greater whole. In other words, there could be some larger meaning behind why and how the universe is unfolding and operating. From this perspective, our "smaller" human meaning is like a fractal of this larger meaning, in the way that many spiritual traditions teach: "As above, so below." The question becomes, what is this larger meaning? Or put more broadly, what is the purpose of our universe and how is this linked to synchronicity?

I suggest that if the underlying structure of the universe is a fundamental whole that is infused with meaning, then it is possible that synchronicity is the bridge between larger universal meaning and our individual human meaning. Synchronicity is a way for the universe to reach its hand toward us while we reach toward it, similar to Michelangelo's painting of God and Adam in the Sistine Chapel (see Figure 3). In other words, synchronicity is a bridge between God and Adam's fingers. Synchronicity is a technique that the universe uses to remind us of this interconnection and of how the meaning that we see in our individual lives is linked in some mysterious way to the larger universal meaning.

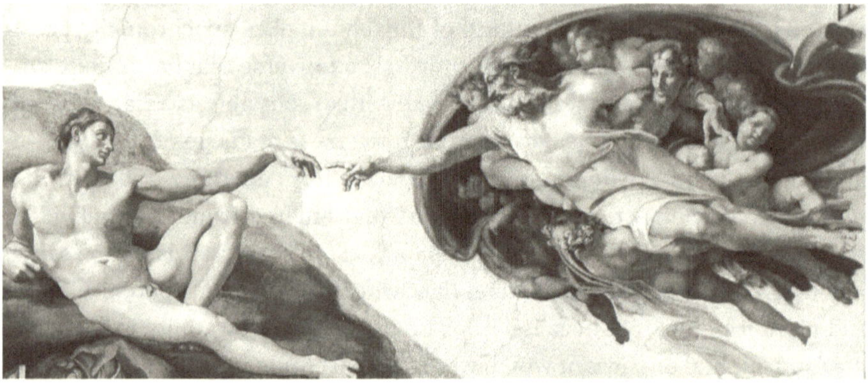

Figure 3: "The Creation of Adam" by Michelangelo in the Sistine Chapel.
Photo by Calvin Craig on Unsplash.

As a researcher and scholar, I am cognizant of the fact that thus far I have offered very few academic sources to support my ideas. On the one hand, this was intentional, as I view my personal experience of meaningful coincidence as being a valid source of evidence in and of itself. That said, I can't help but briefly share some theoretical perspectives within physics, psychology, mathematics, cosmology, and other disciplines that provide potential support for my ideas. The first theory, *panpsychism*, suggests that mind, or consciousness, is present in matter (Koch, 2014). For example, Integrated Information Theory (IIT) uses mathematical equations to suggest that a system's level of consciousness is the result of how much information it is able to integrate (Oizumi et al., 2014). Therefore, according to IIT, any system that possesses a non-zero amount of integrated information experiences some form of consciousness. Thus, in theory, a rock could possess some sort of consciousness, although it would be vastly different from the type of consciousness that we experience. *Idealist* theories suggest that consciousness is more fundamental than matter in terms of making up the basic structure of our reality (Kastrup, 2018). In other words, consciousness "comes first" and somehow leads to matter. From these perspectives, the deepest structure of reality is, in a sense, a Big Mind that is producing the physical reality that surrounds us. *Monist* theories take a similar approach by suggesting that there is some sort of undivided wholeness that is at the root of our reality and from which both mind and matter emerge, however this undivided wholeness is not necessarily conscious or consciousness (Silberstein & Chemero, 2015). For example, dual-aspect monism (Atmanspacher & Rickles, 2022) suggests that the way an object or organism experiences the neutral domain of reality depends on its unique situation and the ways that it generates knowledge. For humans, this means that we perceive matter as existing outside of us, and consciousness as existing in our minds, because of our unique sense organs that "force" us to experience mind and matter in this separate, dualistic way.

Thousands of pages have been written on each of these theoretical perspectives, and while I am only giving them a cursory glance here, I don't want to give the impression that these theories are not well-researched or well-founded. These perspectives are being proposed by thought leaders in a variety of fields, and several have a solid backbone in mathematics and physics. Of course, it is possible that these theories are inaccurate or incorrect. Indeed, there are several theoretical perspectives which suggest that coincidences and synchronicities are simply cognitive strategies that humans use to create meaning in their lives (Colman, 2011), to solve problems (Williams, 2010), and/or to understand events that seem to lack causality (Johansen & Osman, 2015). Along similar lines, some statisticians argue that with a large enough sample, unlikely events randomly happen (Hand, 2015). In other words, synchronicities are nothing more than chance occurrences in the exceptionally large database of the human population, which humans sometimes (erroneously) perceive as being meaningful (see Sawyer, 2018 for a counterpoint to this perspective).

In my view, one of the reasons that the topic of synchronicity has largely remained on the fringes of mainstream science is that modern science is holding onto a materialist perspective that blinds it to other possibilities.

Materialism (or physicalism) suggests that matter is the fundamental structure of reality (Beauregard et al., 2018), and it is a perspective that has brought us many technological advances. But research and theory from a variety of disciplines, such as parapsychology (Cardeña, 2018) and consciousness studies (Wahbeh et al., 2022) suggest that materialism provides a limited view that needs to be expanded by taking consciousness and/or some sort of undivided wholeness more seriously as being an important facet of reality.

Conclusion

To summarize, my view on synchronicity and meaningful coincidence is that they involve cognitive and metaphysical processes that operate simultaneously and synergistically. To go back to the personal examples that I shared at the beginning of this essay, cognitively, my experiences helped create meaning around attending both retreats and around my professional journey of exploring synchronicity as an academic topic. My synchronistic experiences gave me a sense that I was being guided, which comforted me when making the difficult financial decisions that I had to make about attending the retreats as well as decisions about the research topics that I pursue professionally. And, looking back on my experience, I've created an internal narrative structure that it was indeed the right decision for me to attend the retreats and to study synchronicity, which creates further meaning and helps me trust my intuition moving forward.

Metaphysically, I believe that I was in communication and co-creation with some fundamental aspect of reality that was giving me clues not only to guide my life path, but to also show that it (the undivided whole) exists. I'm not referring to what many think of as "God" (i.e., a divine being who is sitting on a cloud while orchestrating the universe). But it is, in a sense, a version of where I believe many ideas about God come from. What I'm referring to is an undivided whole that perhaps has some elements of what we might refer to as consciousness, although another way to think of this consciousness might be as "information." In other words, the undivided whole includes an "information layer" that underlies our experience of reality, and at times, this information pops through the cracks of our ordinary consciousness to guide us and to hint at its own existence. We might think of it as similar to the binary information that underlies computer code. We rarely see the binary code (unless we are a computer programmer) but it is a crucial source of information that creates entire worlds for us to perceive and participate in. Importantly, this information layer is not dead or inert. Similar to the electricity that brings binary code to life in our computer, the information layer is animated by a force or energy that is, in some sense, motivated to co-create with us. This is perhaps similar to the concepts of Shakti and Shiva in Tantric cosmology, where Shiva represents the constitutive elements of the universe, and Shakti is the dynamic energy that brings these elements to life (Temple Purohit, 2022). In my experience, the undivided whole of which this information layer is a part enjoys playing and co-creating with us, when we pay attention to it and acknowledge the possibility that it exists.

I try to keep an open mind to both materialist and post-materialist perspectives on synchronicity and meaningful coincidence, and I think that both perspectives can be co-occurring at the same time. I think that dialogue with those who have different views than ours is important, and that one of the highest forms of wisdom is the ability to hold paradox. Perhaps we can hold materialist and post-materialist views simultaneously and see what might emerge. Perhaps meaningful coincidences are a by-product of my cognitive thought processes and, at the same time, are also a clue about the fundamental nature of reality. What might happen if we allow ourselves to relax into both possibilities being true, rather than trying to immediately prove or defend one or the other? Perhaps, by resting into the field of not-knowing, the answer will, at some point, become apparent.

References

Atmanspacher, H., & Rickles, D. (2022). *Dual-aspect monism and the deep structure of meaning.* Routledge

Beauregard, M., Trent, N. L., & Schwartz, G. E. (2018). Toward a postmaterialist psychology: Theory, research, and applications. *New Ideas in Psychology, 50,* 21–33.

Beitman, B. (2022). *Meaningful coincidences: How and why synchronicity and serendipity happen.* Simon and Schuster.

Butzer, B. (2021). Does synchronicity point us towards the fundamental nature of consciousness?: An exploration of psychology, ontology, and research prospects. *Journal of Consciousness Studies, 28*(3–4), 29–54.

Cardeña, E. (2018). The experimental evidence for parapsychological phenomena: A review. *American Psychologist, 73*(5), 663.

Colman, W. (2011). Synchronicity and the meaning-making psyche. *Journal of AnalyticalPsychology, 56*(4), 471–91.

Hand, D. J. (2015). *The improbability principle: Why coincidences, miracles, and rare events happen every day.* Transworld Publishers.

Hocoy, D. (2012). Sixty years later: The enduring allure of synchronicity. *Journal of Humanistic Psychology, 52*(4), 467–478.

Johansen, M. K., & Osman, M. (2015). Coincidences: A fundamental consequence of rational cognition. *New Ideas in Psychology, 39,* 34–44.

Jung, C. G. (1973). *Synchronicity: An acausal connecting principle.* Princeton/Bollingen.

Kastrup, B. (2018). The universe in consciousness. *Journal of Consciousness Studies, 25*(5–6), 125–155.

Koch, C. (2014). Ubiquitous minds. *Scientific American Mind, 25*(1), 26–29.

Oizumi, M., Albantakis, L., & Tononi, G. (2014). From the phenomenology to the mechanisms of consciousness: Integrated information theory 3.0. *PLoS computational biology, 10*(5), e1003588.

Sawyer, K. (2018). Synchronicity and the search for significance. *Significance, 15*(6), 34-37.

Schwartz, G. E. (2017). *Super synchronicity: Where science and spirit meet.* Param Media.

Silberstein, M. & Chemero, A. (2015). Extending neutral monism to the hard problem. *Journal of Consciousness Studies, 22*(3–4), 181–194.

Wahbeh, H., Radin, D., Cannard, C., & Delorme, A. (2022). What if consciousness is not an emergent property of the brain? Observational and empirical challenges to materialistic models. *Frontiers in Psychology,* 5596.

Williams, G. (2010). *Demystifying meaningful coincidences (synchronicities): The evolving self, the personal unconscious, and the creative process.* Jason Aronson.

Chapter Thirteen

Synchronicity and the Law of Attraction

DAVID LORIMER

David Lorimer is a visionary polymath, poet and spiritual activist who is Global Ambassador and Programme Director of the Scientific and Medical Network, Editor of *Paradigm Explorer* and Co-Chair of the Galileo Commission. His most recent books are *A Quest for Wisdom* and his collection of poems *Better Light a Candle.*

The overall number of minds is just one. The multiplicity is only apparent, in truth there is only one mind.
Erwin Schrödinger

Through my study of late 19th century New Thought, I have come to agree with the above statement made by a pioneering physicist who won the Nobel Prize. Interestingly, his father was a professor of comparative religion, so Schrödinger had early access to the Vedic texts which subsequently informed his worldview.

One striking synchronistic experience occurred at the funeral of my brother Andrew in the Philippines in November 2017. In the middle of the service, a cat came up the aisle, passed my brother's coffin then came over to my sister-in-law Annabelle before walking back out. My brother had many cats and even referred to himself on Facebook as 'The Baguio Cat Man.' The people associated with the church had never seen a cat come into the building before, so the timing and significance were remarkable.

My other experiences of synchronicity are perhaps not classic, but they were certainly meaningful and significant, representing two key turning points in my life. In December 1979, I was teaching at Fettes College in Edinburgh and had no immediate thoughts of moving on. I had many friends in the city, and my home was only an hour away in Fife. I went to Cupar to have lunch with my French poetry tutor Ian Higgins. As I was early, I went into

John Menzies stationers where I bought a copy of the Times Educational Supplement. I casually turned to the vacancies section where I saw an announcement—Winchester College, French and German. As soon as I set eyes on this announcement, I knew I was going to get the position. I put in my application and took the overnight train to London and then on to Winchester. I walked down from the station to Winchester College and went into the Cathedral on the way, where the organist was playing Bach's D major (BWV 533) fugue. I have a great love for Bach's organ works, so I regarded this as a good omen. There followed an interview with the brilliant Head Master John Thorn and a grilling from the head of modern languages, John Surry. I returned to Edinburgh, and just over a week later, I found an envelope lying in my staff post box marked Winchester College—I vividly remember opening the letter in the staff passageway. It was from John Thorn, saying that they were looking forward to welcoming me to the College in September—so my original intuition had been spot on. Years later, there was an interesting sequel connected with David Conner, who was appointed as Senior Chaplain at the same time as me and became a great friend. More than 30 years later, I went to visit him in Windsor where he had been Dean since 1999. At the end of a very congenial evening, he escorted me to the gate, and when we passed St George's Chapel, I heard the very same piece of Bach being played by the organist—I had not heard it since that first occasion.

My second experience concerned moving on from Winchester. In 1983, I received an invitation out of the blue from George Blaker to join the Scientific and Medical Network. I duly sent off my subscription and in his reply George invited me over for lunch. I discovered that we were fellow old Etonians, and he took me down to his local pub, The Crown in Capel. After about 10 minutes, he said to me that I was just the kind of person he was looking for to take over the Network. I immediately knew intuitively that this was my next step in life, and I am still working with the Network 40 years later.

I interpret these experiences as part of a deeper soul blueprint, as reported in *Testimony of Light* by Helen Greaves, a book I read in 1975. Through contact with her late friend Frances Banks, it was explained that we each have an intended life blueprint before we incarnate, and our task is to tune into this and to follow it. I regard these two incidents as occasions when I had a strong sense of my blueprint as it was unfolding in time. In 2010, I went back to Eton College for a 40th anniversary dinner for those who left the school in 1970. While walking across School Yard from the Chapel, I had a very distinct sense of inner affirmation in having followed this blueprint, which I still feel vividly present—the sense of being meaningfully on track.

In 1990, I published a book called *Whole in One—the Near-Death Experience and the Ethic of Interconnectedness.* In it, I focused on the moral significance of the life review, with case history examples not only from NDEs, but also from post-mortem communications. In the deeper such experiences, events are relived multi-dimensionally whereby subjects actually experience themselves the feelings and emotions—both positive and negative—of other people involved, and even the indirect ripple effects on third parties. The keys are responsibility and love; in this respect, Peter Deunov summarizes the message by saying that when we return to the invisible world, we will be examined on how we have applied the law of Love. The centrality of love in human life present throughout the NDE literature is further highlighted by a recent compilation of NDE philosophy by Douglas Hodgson—*Spiritual Revelations from beyond the Veil.* My overall argument is that if we knew that a life review awaits us all, the world would be a very different place.

At the time, my model was based on the work of David Bohm's Implicate Order (Bohm, 1980) whereby our individual consciousness is a relatively autonomous sub-totality abstracted from cosmic consciousness or universal mind (whereas at the time my understanding was that we were all part of a whole, I now think that we *are* the whole as individual expressions). At the explicate level of ordinary waking reality, it looks as if our individual consciousnesses are separate, while at the implicate level, there is what Bohm called unbroken wholeness of consciousness in flowing movement. This is directly experienced in mystical and unitive states. The phenomenology of the life review indicates just such a deep level of interconnectedness of mind and consciousness, otherwise the subjective reliving of the experience of others would be impossible in principle. On this basis, I postulate our capacity for what I called 'empathetic resonance' and its ethical implication that we should conduct ourselves in accordance with the Golden Rule. At a deeper level, we are one another, so what we do to each other we are ultimately doing to ourselves. We may only realize this fully in the life review. This entails what I call an ethic of interconnectedness. Since that time, a good deal more evidence has come forward from various disciplines of an underlying unity and consequent interconnectedness with a philosophical emphasis on process and relationality.

My views on consciousness and the interconnected nature of reality have been influenced by New Thought School of thinkers from the late 19th century as well as transcendentalists like Ralph Waldo Emerson—see his essay *The Oversoul.* Contrary to emerging positivist trends in philosophy and psychology, these thinkers postulated the existence of a Universal Divine Mind, as referred to above. Here is the formulation from Charles Haanel (1866–1949), taken from Lesson One of his 1912 publication *The Master Key System:*

'There is one Principle or Consciousness pervading the entire universe. As this Consciousness is omnipresent it must be present within every individual; each individual must be a manifestation of that Omniscient, Omnipotent and Omnipresent Consciousness. Your consciousness is (therefore) identical with the Universal Consciousness, all mind is one mind. Each individual is the individualization of the Universal, the Cosmic Mind.'

While Freud and Jung popularized the concept of the subconscious and unconscious, New Thought thinkers added what they called the Superconscious Subjective Universal Mind. As Haanel put it: 'Through the [superconscious] we are connected with the Universal Mind and brought into relation with the Infinite constructive forces of the universe.' In his bestseller, *In Tune with the Infinite,* Ralph Waldo Trine (1866–1958) emphasized the importance of aligning ourselves: 'The great and central fact in human life, in your life and mine, is the coming into conscious realization of our oneness with this Infinite Life, and the opening of ourselves fully to this divine inflow. What does this mean? That we are recognizing our true identity, that we are bringing our lives into harmony with the same great laws and forces as have all the prophets, seers and sages in the world's history.' (Trine, 1920) Similarly, Thomas Troward wrote of the individual mind as a 'Centre of Divine operation' where our task is self-recognition, coming to know that what can say 'I Am' in me is the same as what can say 'I Am' in you. This is what the ancient Greeks called *gnosis,* direct realization that we are the same ultimate Divine Consciousness, the One. Our true common identity is universal, as Ramana Maharshi also realized: self is ultimately and essentially Self.

A further premise is that thoughts are creative and causal, while outer conditions reflect the effects or manifestations of these thoughts. Here is Haanel: 'Thought is energy, active thought is active energy, concentrated thought is concentrated energy ... thoughts are causes, outer conditions are effects. The world without is a reflection of the world within. A majority of mankind lives in the world without; few have found the world within and yet it is the world within that makes the world without.' (Haanel, 1912) The same is still true in our time. Thomas Troward therefore asserts that the creative order is from [inner] states to [outer] conditions, but a third-person perspective inverts this order and seeks to create states from conditions.

I sometimes give the example of my grandfather, the Scottish architect Sir Robert Lorimer (1864–1929). I am writing this piece at a bureau that he designed, and sitting on a chair also designed by him. These pieces of furniture were manifestations of his creative imagination, as were his great works the Thistle Chapel, Ardkinglas Castle and The Scottish National War Memorial. All these began as ideas that became sketches, designs and plans that were subsequently constructed in the physical dimension. If you look around at any artefact, you will realize that it began as an idea in someone's imagination.

A further consequence is also elaborated by Haanel on the basis that like attracts like: 'Thought is a creative energy, and will automatically correlate with its object and bring it into manifestation.' Troward sees this in terms of the law of growth and the centrality of desire as a dynamic energy. For him, the Law of Growth is a universal creative force throughout nature, whether on the inner mental or outer physical plane (Troward 2007, p. 23). In each domain, the seed is planted that leads to a harvest as cause is related to effect. The seed attracts to itself the necessary elements for its growth. In the same way, he argues, an idea planted in the subconscious mind begins to attract to itself the necessary means for its fulfilment. This includes people, events and circumstances that seem to conspire to bring things about. Readers can appreciate that there seems to be a connection between the creative nature of thought and the law of attraction as manifested in synchronicity, although this does not seem to be consciously the case in my own experiences, even if they are examples of an acausal connecting principle.

As Ralph Waldo Trine put it: 'Each is building their own world. We both build from within and we attract from without. Thought is the force with which we build, for thoughts are forces. Like builds like and like attracts like.' (Trine, 1920, p. 1) Life doesn't just happen to us—it responds to our dominant thoughts and feelings as we are all immersed in an implicate, interactive, conscious energy field, as this quote from Earl Nightingale suggests: 'Our attitude to life determines life's attitude towards us.'

For Troward, desire is the creative power, the force behind all things, the moving principle of the Universe, the innermost center of all Life, the cause of all feeling and action. (Troward, 1921, p. 55) Desire is the mind seeking to manifest itself in some form which as yet exists only in thought, it is the principle of creation whereby the creative process is identical at all levels in nature, inner and outer. Plant, grow, ripen, harvest. This is the secret. We as individuals are creative centers of our own world. Life is being, an experience of states of consciousness, with a correspondence between inner states and outward conditions. Hence, as already stated, states of consciousness are regarded as causal and the corresponding outer conditions the effects. Our own creative thought, reflecting the inherent creativity and generativity of the Universal Mind, has both a *formative and attractive power,* and it is important to highlight that our ultimate desire is for unitive experience of 'Being'—which we already essentially are—rather than 'having' in terms of acquiring material possessions. For mystically-oriented Christians, this is the Inner Christ as omnipresent essence.

All this implies that Consciousness as ontologically and epistemologically fundamental (I prefer this term to primary, as the Ground of Being gives rise to energy/life/consciousness and form as vehicles of expression in space-time). This corresponds to the relationship between essence and existence, absolute and relative, being and becoming (prior and consequential for Whitehead). One becomes two when the unmanifest becomes manifest in form, the individual is the reciprocal complement of the universal, as Thomas Troward argued: the universal exists within each of us and we exist within the universal. Or, as Plotinus put it, we are within a reality that is also within us. This entails a Macrocosmic Universal Mind of which we are all intrinsically interconnected microcosms: there is One Mind and One Life. For me, this implies the Golden Rule, or what I call an ethic of interconnectedness, which I also derive from the life review, as articulated above.

In the 1990s, we in the Scientific and Medical Network held a number of working group meetings on what we called 'participatory reality,' inspired by, among others, Richard Tarnas' *The Passion of the Western Mind* and *The Participatory Mind* by Henryk Skolimowski. Tarnas explains his view in the Foreword (Tarnas, 1991, pp. 434 ff.) whereby 'the essential reality of nature is not separate, self-contained, and complete in itself, so that the human mind can examine it "objectively" and register it from without. Rather, nature's unfolding truth emerges only with the active participation of the human mind.' Descartes was wrong when he split off mind from matter, inner from outer when they are in fact entangled, as synchronicity demonstrates. Both

Rudolf Steiner and Owen Barfield understand this as an evolutionary process whereby the original participatory consciousness is reduced/contracted to non-participation and indeed alienation in the scientific revolution. In the wake of Goethe, we are now moving towards renewed participation in the universal (Barfield, 2024, p. 134). This overall movement corresponds to the process thinking of Whitehead and the primacy of relationality in Iain McGilchrist; also wave-particle complementarity in physics. Instances of synchronicity instantiate an epistemology of participation rooted in a unitive ontology.

Synchronicities are by definition meaningful coincidences of interconnectedness and indeed of flow. They invite an entry into the symbolic, into the depths, into the unconscious. Spirituality and meaning are both to be found in connection, in terms of relationships and spirituality. The dominance of chance, randomness and natural selection in evolutionary biology creates a resistance to recognizing meaningful coincidences, reinforced by sharp divisions between subjective and objective that are now coming under scrutiny in the mainstream. A significant new book by Adam Frank, Marcelo Gleiser and Evan Thompson is entitled *The Blind Spot*—why science cannot ignore human experience. It calls for a revolution in the scientific worldview whereby humanity's lived experience becomes an inescapable part of our search for objective truth. In this view, science becomes a highly refined, constantly evolving form of human experience.

In answering the question relating to what kind of universe makes synchronicity possible, our worldview has to evolve beyond materialist orthodoxy that not only dismisses synchronicity but other well-attested parapsychological phenomena as impossible in principle on the basis that these things can't happen, so they don't. However, Alfred Russel Wallace pointed out 150 years ago that there are no impossible facts and that if your theory fails to account for the facts in an adequate fashion, then it is the theory that needs an upgrade. The emerging worldview is one in which synchronicity becomes logically possible and indeed plausible. It is not based on sharp divisions between inner and outer, subjective and objective, mind and matter, qualitative and quantitative but rather on such principles as fields, unity, entanglement, nonlocality, complex systems, holograms, fractals, autopoiesis, distributed intelligence, inherent creativity, interconnected ecosystems, interactive participation, relationality and process. Such an outlook will enable us to recover a sense of the lost Renaissance *Anima Mundi* to realize anew our embeddedness in Earth systems and as 'interbeing' in each other so that we can act accordingly.

References

Barfield, Owen (2023). *The Riddle of the Sphinx.* Barfield Press.

Frank, Adam, Gleiser, Marcelo, Thompson, Evan (2024). *The Blind Spot.* MIT Press.

Greaves, Helen (1969). *Testimony of Light.* CFPSS Press.

Haanel, Charles (1912, 2007). *The Master Key System.* Ishtar Publishing.

Lorimer, David (2017, 1990). *Resonant Mind.* White Crow.

Trine, Ralph Waldo (1920). *In Tune with the Infinite.* G. Bell.

Tarnas, Richard (1991). *The Passion of the Western Mind.* Ballantine Books.

Troward, Thomas (1921). *The Hidden Power.* R. M. McBride.

Troward, Thomas (2007). *The Wisdom of Thomas Troward.* Wilder Publications.

Chapter Fourteen

Evolving through Synchronicity in an Archetypal and Symbiotic Universe

MONICA BRYANT, MA

Monica Bryant, Evolutionary consultant and coach, has provided archetypal perspectives to support soul evolution, since 1994. She is an independent trans-disciplinary researcher with an ecological and holistic worldview. Monica's MA in Myth, Cosmology and the Sacred was on transformative learning, synchronicity and transpersonal research methodologies. At the University of Sussex (1980s) she taught Culture, Health and Healing, Goethean Science and The New Microbiology. She pioneered symbiosis-restoring approaches for health, lecturing internationally. Monica was Director of her health company for 30 years. Memberships: Scientific and Medical Network since 1983; Association of the Emerson Trust; Academy of Professional Dialogue; and transpersonal communities. www.evolutionary.co.uk

Synchronicity has been an important, integrated part of my life, like a constant companion. My experiences have varied in magnitude in their subjective effects, from small daily delights to major leaps on my life path that have brought new projects and opportunities to enlarge my horizons. The purer my soul intention to be of service to the greater whole, the more life-changing synchronicities have occurred, with portals opening for new archetypal expressions in my life.

My synchronicity experiences have been enriched through the symbolic meanings of birds, wildlife, insects, numbers and colours, which have acted as reassuring signs. In the 1970s, when I began to meditate and study astrological cycles, the frequency of synchronicities increased, providing an endless source of meaning, self-knowledge and a sense of participation in the cosmos. Two significant examples, where I experienced chains of multiple coincidences with great transformative impact, follow.

California calling—expanding my conscious awareness

The first was a dramatic synchronicity that accelerated my personal evolution. It concerned an inner calling to go to California, following my university graduation in 1982. As I did not know anyone in the USA, I went to the Mind, Body, Spirit Festival office in London, hoping to gather any helpful contacts. Shortly after arriving there, I was offered a temporary job in San Francisco. Although surprised, I immediately had an inner 'yes' and accepted. A few weeks later, I was catapulted into the hub of the human and planetary potential movements. Mill Valley in Marin County became my home, surrounded by creative pioneers, healers and spiritual retreats. My office was near to the Golden Gate Bridge; coincidentally, it was next to the publisher of *New Realities,* who was the husband of Jean Shinoda Bolen, author of *The Tao of Psychology: Synchronicity and the Self.* Discovering a wide range of psychospiritual therapies, approaches to holistic health and the emergence of leading-edge, new paradigm thinkers was uplifting. I met many inspirational authors, including Willis Harman, Buckminster Fuller, Fritjof Capra, Richard Grossinger and Barbara Marx Hubbard. As a guest, I was invited to the dynamic Esalen Institute and connected with transpersonal psychology pioneers such as Michael Murphy, Frances Vaughn and Dane Rudhyar. Social symbiotic alliances were formed to serve the higher purpose of 'saving the world'. Working within this holistic community that encouraged networking, frequent synchronicities occurred with interconnections touching every aspect of life.

My 'California calling' synchronicity experience was personally meaningful because, following my intention to find some Californian contacts, I suddenly found myself employed to make connections with hundreds of extraordinary people. Out of the blue, like a cosmic seed moment, a portal opened to a new, expanded and multidimensional life, harmoniously aligned with a deeper sacred order, as depicted in figure 1. I was immersed in the field of consciousness research, involving a multitude of expressions for mind, body and spirit. It was a time filled with unitive peak experiences, which transformed my consciousness and continues to inform my awareness today. New synchronicities are still emerging across space-time with more subtle iterations of the same people and intellectual themes that I encountered then—revealing a meaningful fractal patterning in my life.

From 1982 to 1983, I also worked with Peter Russell to promote *The Awakening Earth* in the UK then *The Global Brain* in the USA, based on Teilhard de Chardin's view and a Gaian understanding of Mother Earth as a self-regulating, conscious superorganism. Russell (1982, p. 197) envisioned an increase in supportive synchronicities as an expression of a healthy 'high-synergy society', which aptly summarises my California experience.

Figure 1. Life path portal to the multidimensional, sacred and evolving cosmos.
(Retrieved 10 April 2023. Royalty-free image. https://www.dreamstime.com/web-image 198232135#res12876883)

An evolutionary group—multiple synchronicity experiences

The second example occurred from 1984 to 1994 when I was part of a synchronicity-led international group of organisational transformation consultants who were creative and inner-directed. Our group process became an experiential study where we observed consciousness unfolding, emergent phenomena and archetypal dynamics. As we witnessed what happened in our 'being' together rather than 'doing', what emerged out of unknowing, nothingness or emptiness were synchronicities, which we called 'surprises'. To name our group, I proposed the Surprise Being, because the synchronistic surprises opened us up to evolve individually and collectively. As a group, there was no defined outer purpose or usefulness for us to continue meeting. Therefore, synchronicities created shared meaning and an internal generative process that ensured our survival as a group by planning to meet again.

During this intensive group process, we went from the edge of chaos to reordering into new developmental stages. Together we were contemplative and embraced the transpersonal dimension. In the mystery and uncertainty, I found that the potential self-organising principles were revealed through various phenomena, given time, often following spontaneous silences. I deepened into the flow where polarities were transcended and a third new element creatively emerged. The inner experience often felt like a symbolic death and rebirth as we evolved to a new level of wholeness. In the Surprise Being, we experienced a deepening interiority in our evolutionary process as we created a loving community that felt like a palpable group soul or a coherent organising field of consciousness.

We had clusters of synchronicities through sharing our dreams over breakfast. Coincidentally we frequently wore the same or similar colours. In sharing our personal lives the coincidences were deeply meaningful and sometimes highly amusing. At airports, we spontaneously coalesced to meet up without prior plans. We were often playful and the various meanings from our synchronistic experiences strengthened our bonds. Each time we gathered to meet, a coincidental theme would soon become apparent; for instance, on one occasion, several of us had just experienced a death in the family, which became our shared focus. In between our meetings, I started a newsletter called the *Surprise Scanner* to continue monitoring any emergent phenomena in our lives and significant synchronicities. As we lived in different countries, this helped us to maintain our group intimacy and explore our interconnectedness beyond space-time.

In the Surprise Being, as we sat in a circle, I watched the dynamic interplay of archetypes and hidden geometries unfolding into mandala-shaped patterns, like looking into a kaleidoscope, as envisaged in figure 2. Self-organising clusters and constellations emerged, receded and vanished in our group as expressed through our interpersonal dynamics, ideas and feeling states. The cosmos flowed through us in a dance of energies, polarities and archetypal expressions which morphed in a variety of ways. Then, when we reached a place of resolution from our different positions, we would experience an equilibrium and flow through our group as a complete circle. This was a whole where our diversity was contained within a unity. Our evolutionary group process created a kaleidoscope of cycles within cycles, ever-changing and transforming, yet sustaining the dynamic harmony essential to the evolution of life.

Figure 2. Cosmic mosaic-like kaleidoscope unfolding mandalic patterns.
(Retrieved 10 April 2023. Royalty-free image. https://www.dreamstime.com/stock-photo-image 48105707#res12876883)

The personal significance of my 'evolutionary group' synchronicity experiences were many. I learned to trust that synchronicities would happen at the right time allowing creative emergence to occur. In the Surprise Being we provided a safe, loving container with deep listening on all levels, which enabled us to express what was emerging out of our souls and give birth to our new selves. Conditions conducive to synchronistic phenomena may include peaceful states, through quietening the mind to access deeper levels of consciousness. Various forms of meditation have been found to increase the frequency and influence of synchronicities (Russell, 1982, p. 195).

To summarise our experiences of evolving together, I described the emergence of a new form of 'evolutionary group' (Bryant, 1999). To define the criteria of an evolutionary group, I integrated the following interpretive lenses: David Bohm's group dialogue (Factor, 1985), Ken Wilber's evolutionary transpersonal theory, James Hillman's archetypal psychology, Rupert Sheldrake's morphogenetic fields, Johann Wolfgang von Goethe's participatory and phenomenological observation, complexity and chaos theory, and process-oriented community building. This profound group process formed my evolutionary consulting and coaching approach, as I could not find other models or techniques that could match the immediacy, aliveness and transformative capacity that we shared. Synchronicity experiences continue to guide and enliven my intuitive research.

Surprise, emergence and new levels of wholeness

These two synchronicity examples are significant for me today because they were part of shaping my understanding of consciousness and the interconnected nature of reality. Synchronicity creates a crack in time to let the cosmos in by the opening of our hearts and minds. A meaningful coincidence can initiate surprise, or similar states of bewilderment and awe, that bring us into the present moment to shift habitual patterns and initiate changes. Surprises have a spectrum of intensity, from feeling stunned, shocked or startled to more gentle states of wonder and curiosity, opening us to new awareness. In the holistic theory of complexity, as applied to the psyche, surprises are a form of novelty in living systems. Synchronicity 'is not based on cause and effect rationality but reveals itself through the emergence of self-organising features that evoke a feeling of surprise from curiously mild serendipities to stunning coincidence of significance' (Cambray, 2009, p. 107).

In the Surprise Being, through our cycles of death and rebirth experiences together, we were born as a social holon. The term holon (a whole/part) was coined by Arthur Koestler (1967, p. 48) to describe how every part belongs to a whole and every whole is a part of another whole. Holons have a basic polarity between—the integrative tendency of partness and the self-assertive tendency of wholeness (Koestler, 1967, p. 56). Holons are embedded—one within another—like a series of nested concentric circles to form a holarchy which can be any living system, such as an organism or an individual. In evolution, each holon aspires to the next level of wholeness, through inclusion and then moving beyond the previous holon. As a group, we had both incremental and major evolutionary steps, thereby creating greater levels of holarchic depth. Synchronicities acted as an acausal catalyst for emergence, revealing our universe to be self-organising, intelligent and purposeful.

Evolutionary endosymbiosis, holism and synchronicity

Holistic and ecological thought is central to my worldview and to understanding how ecosystems are self-regulatory and evolutionary through the role of symbiosis. The concepts of symbiosis and synchronicity share the same etymological Greek root, *sym,* which means 'together': its pairing with *bios,* 'life', means 'life together', while *syn,* joined with *chronos,* means 'time together'. Symbiosis is an ecological, relational and participatory concept that is appropriate in the context of synchronicity, which bridges inner and outer, psyche and matter.

Our deep interconnectedness with the microbial world is integral to our evolution through endosymbiosis, which is a symbiotic relationship where one organism lives inside another (Bryant, 1986). This involves the incorporation of unicellular organisms to form inner organelles, such as mitochondria, within multicellular host cells. A mutually beneficial relationship of endosymbiosis is where a part becomes included in a new whole or holon to form a holarchy of organisms enabling greater complexity and evolutionary creativity.

An ecological view of the psyche may be seen as 'an ecosystem of communicating conscious agencies' (Kastrup, 2021, p. 29). The integration of synchronicity experiences within the psyche might be analogous to an endosymbiosis with similar evolutionary dynamics, by forming an intrapsychic symbiosis. For example, by reflecting on my synchronicity experiences, they have continued to be a powerhouse of new insights and meanings; metaphorical mitochondria as integrated organelles in my psyche. This ecological and holarchic perspective is aligned with Carl Jung's quest for wholeness. Main (2019, p. 14) elucidated Jung's position of ontological holism. He also demonstrated that Jung's 'concept of synchronicity is itself deeply holistic' (Main, 2019, p. 19).

Synchronistic experiences: the evolutionary and archetypal nature of reality

Consciousness for me is a cosmic unity—the fundamental ground of all Being or the *Unus mundus* (One world) out of which life manifests through organising principles or ordering patterns. These could be described as fields such as morphogenetic fields, archetypes, souls, holons or strange attractors. The undivided Oneness becomes separated into two, such as the dynamic and complementary opposites of yin and yang in Taoism. The four elements of air, water, earth and fire—and Jung's psychological typology of thinking, feeling, sensation and intuition—are timeless ways to understand reality.

Much of my inspiration comes from Goethe's intuitive and holistic mode of perception of archetypal forms in nature and their metamorphoses. His catalysing principle of polarity enables opposites to create a great variety of phenomena in natural, personal and cultural life. For Goethe, higher development was attained through an evolutionary energy striving towards synthesis and wholeness in all aspects of life. Goethe's view was evolutionary through a process of intensification and progression *(steigerung)* of the two poles, creating an unexpected and higher 'third thing'. The process of becoming is an ongoing evolutionary imperative that moves towards new holons of greater levels of depth.

The evolutionary and archetypal nature of synchronicity I view primarily within a Jungian epistemology. I perceive the nature of reality as an interplay and confluence of archetypes in endless creative dynamics. Archetypes emerge from the collective unconscious and manifest through physical and psychic phenomena in various domains of life to bring forth a richness of qualities, nuances and symbolic meanings.

I have a deep connection to the cyclic nature of reality through the yearly seasons. The planetary archetypes offer me the most direct way to experience the cyclic qualities of time and their potential meanings for my evolutionary journey in a cosmic context. Of the countless archetypes, the time-honoured myths of goddesses and gods have deep resonance patterns. Based on extensive archetypal research of planetary correlations, human experiences and depth psychology, Tarnas (2006) developed an archetypal cosmology which illustrates synchronistic correspondences.

Synchronicity, unfolding meaning and archetypal depth

To unfold meaning from my synchronicity experiences and for them to become fully integrated, embodied and ensouled, I work with symbolic interpretation, contemplation and my body's felt sense to help me gain deep insights. By engaging with the imaginal realm, I observe archetypes and their many psychophysical expressions avoiding literalism, as emphasised by Hillman (1992). This approach is supported by harmonising my right-hemisphere capacity for intuition and imagination with analytic discernment in an inter-hemispheric symbiosis (Bryant, 2021a).

An archetypal reality provides me with countless potential meanings and the perception of levels of archetypal depth (Bryant, 2021b). Direct apprehension and revelation of archetypes can be utterly transformative. Such was my experience while studying the scarab beetle images in Jung's defining example of synchronicity, illustrated in figure 3:

1. Physical scarabaeid beetle that appeared at Jung's window which he passed to his patient coincidentally as she was recalling her dream.
2. Symbol of a golden scarab beetle as jewellery in the patient's dream.
3. Myth of the Egyptian sun-god changing into the scarab which becomes rejuvenated under the morning sun.
4. Spiritual numinous experience of rebirth by resonating with the scarab archetype.

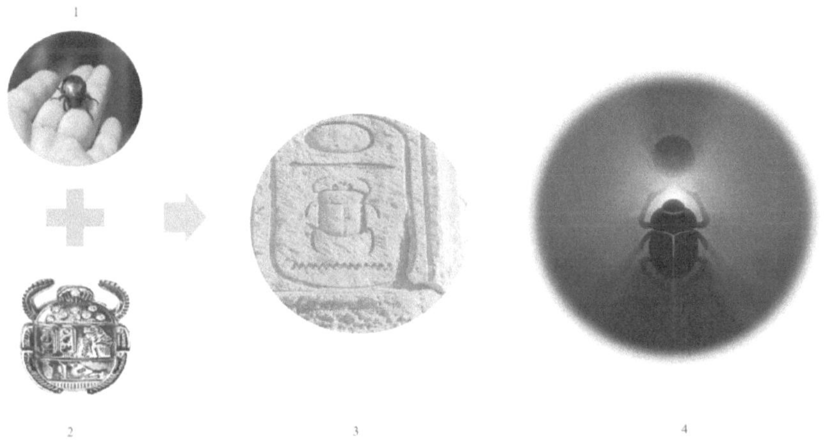

Figure 3. Depth experience of the scarab beetle archetype of rebirth. (Copyright © Bryant 2024).

I experience intuition as closely linked to synchronicity phenomena. Following my intuition as an evolutionary impulse has given me soul guidance with a feeling of 'coming home' or a return to self, which is deeply nourishing. I experience significant synchronicities as a euphoric unification of the inner and outer aspects of my life. Overall, the suffering in life has been balanced by synchronicity experiences to ultimately validate a sense of my being embedded in a relational, loving and meaning-filled cosmos.

Eros enables a symbiotic and pansynchronistic universe

Co-creatively, synchronicities can bring benevolence and expansion through an evolutionary embrace of love (Bryant, 2021a). Jung described the principle of Eros as 'the great binder' (1964, para. 255). It is a primary force of coupling—a 'connecting principle'—which brings inner and outer together in creative syntheses. Eros was equated with the integrative tendency of holons by Koestler (1967, p. 58) and is therefore applicable to both symbiosis and synchronicity. Eros as a co-evolutionary force of attraction, affinity and love may be the convergent force underlying evolution, whether synchronistic (in time) or symbiotic (in space). Might there be a synergy between synchronicity and endosymbiosis in initiating emergence and spontaneous evolution?

Paul Kammerer (1913) discussed the concept of symbiosis and mutual cooperation envisaging homologies in the symbiotic microcosm and in the macrocosm which he called 'pansymbiosis'. Philosophically, Jung viewed Eros as 'pan-Eros who permeates all nature as a creative and procreative force' (Jung, 1964, para. 5). Since the etymology of *pan* means 'all', I propose a new term of 'pansynchronicity', which expresses—all together now—the evolutionary and archetypal interconnections between nature, the psyche and the cosmos.

Conclusion

The subject of synchronicity is multidimensional, prompting an increasing amount of transdisciplinary scholarly literature. For future research, there is value in delving into further historical influences on Jung's and Wolfgang Pauli's thinking that remain implicit and unarticulated. They shared the same German intellectual milieu as other scientists in search of universal laws and looked to the natural sciences for analogies and metaphors which could be applied to the psyche. The history of psychological and biological holism and the German idealist tradition may offer more insights into the nature of consciousness. The Romantic scientists integrated visual skills, art, poetry and reflexivity before the academic disciplines became more specialised. They had the courage to follow their soul's calling and honoured the treasures of their inner lives. This undervalued holistic scientific knowledge is now needed to rebalance our materialistic, ego-based and fragmented world. It is timely to reconnect with lost wisdom and the art of sympathetic resonance to find life's deeper order, meaning and harmony.

For synchronicity in an evolutionary, symbiotic and archetypal worldview, it is important to provide education in the language and dynamics of archetypes. Combined with fostering holistic modes of perception through comparative phenomenological observation and transpersonal ways of knowing. Synchronicity experiences could be included as part of qualitative research methodologies, together with an intuitive, archetypal and self-reflective approach. Such awareness has a role in nurturing individuation in the context of soul-based education and is a transferable skill for life. Synchronicities can connect us with the embracing mother archetype and highlight our interdependence within the divine, symbiotic universe.

References

Bryant, M. (1986). Microbiology at a turning point. *Journal of Alternative Medicine,* March 1986, 4–6.

Bryant, M. (1999). Evolving together: the emergence of a new form of evolutionary group. In H. Jackson (Ed.), *Creating Harmony: Conflict Resolution in Community* (pp. 61–74). Gaia Trust and Permanent Publications.

Bryant, M. (2021a). Intuition as an evolutionary impulse that expands wholeness, wonder and wisdom. In *Anthologi Det Intuitiva* (pp. 75–98). Förlagshuset Siljans Måsar.

Bryant, M. (2021b). Jung's theory of synchronicity, archetypal depth, and an evolving worldview. *The Pari Perspectives Journal, (8),* 109–115.

Cambray, J. (2009). *Synchronicity: Nature and Psyche in an Interconnected Universe.* Texas A&M University Press.

Factor, D. (Ed.). (1985). *Unfolding Meaning: A Weekend of Dialogue with David Bohm.* Foundation House Publications.

Hillman, J. (1992). *Re-Visioning Psychology*. HarperPerennial.

Kammerer, P. (1913). Pansymbiose. *Die Natur Wissenschaften (50),* 1222–1225.

Kastrup, B. (2021). *Decoding Jung's Metaphysics: The Archetypal Semantics of an Experiential Universe.* Iff Books.

Koestler, A. (1967). *The Ghost in the Machine.* Hutchinson.

Jung, C. G. (1964/1970). *The Collected Works. Vol. 10: Civilisation in Transition* (2nd. ed.). (Sir H. Read, M. Fordham & G. Adler eds.; W. McGuire, Exec. ed.; R. F. C. Hull, trans.) Routledge & Kegan Paul.

Main, R. (2019). Synchronicity and holism. In *Analytical Psychology meets Academic Research: Avignon Conference 2018,* 59–74. *Revue de Psychologie Analytique* (Hores série).

Russell, P. (1982). *The Awakening Earth: Our next Evolutionary Leap.* Routledge & Kegan Paul.

Tarnas, R. (2006). *Cosmos and Psyche: Intimations of a New World View.* Viking Penguin.

Chapter Fifteen

Harvesting the Fruits of the Synchronicity Field

VASILEIOS BASIOS, PHD

Dr. Vasileios Basios, a senior researcher at the University of Brussels, specializes in interdisciplinary research on self-organization and emergence in complex systems and the role of information flow and processing. He is passionate about exploring the history of scientific ideas and advocating for a qualitative shift in science towards a post-materialistic view. Basios sees complexity studies as pivotal in this transformation, emphasizing the importance of introducing self-reflection into scientific practice. He aims to contribute insights from complex systems science to foster a resilient network of individuals, groups, and organizations aligned with this vision.

Synchronicities about synchronicity I

Correlations, local, global and non-local, in complex systems, and the mathematics that studies them, have been an important but side-line subject that has accompanied me ever since I can remember my academic life. In parallel and under the radar of institutional orthodoxy, similar questions have driven and guided my other extracurricular studies and research activities. These other studies were on the path that Robert (Bob) Jahn, Brenda Dunne and Roger Nelson paved for us all through the famous experimental era of the Princeton Engineering Anomalies Research Laboratory and the first era of the Global Consciousness Project. In a nutshell, they showed that intention, attention and major shifts in states of consciousness in individuals and groups of people—from small audiences to large populations—leave statistically significant and repeatable weak but persistent traces as deviations from 'pure' randomness produced by their special electronic devices, the 'REG', short for Random Event Generator.

At the time, there was a conceptual gap, at least in my mind, that kept raising an unanswered question in the field. Are these deviations from randomness correlated, and if so, how? Locally, i.e. by a physical field via a causal mechanism? Globally, like the emergent phenomena in the flight of a flock of birds, where they are causally correlated but manifest on much larger scales of time and space than the interactions in the flight of individual birds? Non-locally? As in the well-known a-causal way of the famous entanglement experiments predicted by John Bell? And if so, could we detect them from the existing data? Could we set up and run an experiment that would test for the famous 'Bell-type inequalities' that are so familiar to anyone studying 'entanglement' quantum or not?

Could we, should we, use these REG anomalies to detect extreme changes in consciousness such as dying and end-of-life experiences (terminal lucidity, etc.), out-of-body experiences (OBEs) and even near-death experiences (NDEs)? Would we need a special setup to measure this field, which is acausal and random or rather "synchronic" in nature? Why not link it to collective phenomena even in social animals like bees ... ah, the bees! Really, do non-human animals share the same or different synchronicity field(s) with human animals? ... Anyway, you get the idea, these were the sorts of questions we started to share, mostly in confidence and in duos or trios, as they were probably half-baked ideas, and pondered over with a few good and trusted friends[1].

And then the pandemic struck. Lockdown. After the initial shock, online meetings became more and more frequent. The Galileo Commission's webinars began and flourished in number and quality. You see, the answer to isolation was an almost instinctive reinforcement and acceleration of these online meetings and the formation of many new groups of people talking and being together in this new way, using newly emerging applications and platforms and 'online stuff'. The living network of networks of like-minded organisations, individuals, projects and ideas that the Galileo Commission had envisioned took on an unexpected momentum.

In preparation for one of my own webinars, I was thinking again about how to efficiently detect correlations of REGs: could they be "entangled" in

1 These included Brenda and Jeff Dunne, a physicist, both of the ICRL, the successor to the PEAR lab, Pierfrancesco Moretti, a physicist with two PhDs (!), an NDE(!!) and a specialist in signal processing, our good old mentor Peter Fenwick, the renowned neuropsychiatrist, and Christina Stadlbauer, a visionary biologist and deeply involved mellitologist, bee specialist.

a larger space that would include the noetic and qualia as acausal relations? How can space-time and this very different relational space of qualities be described in a mode à la "unus mundus"? Reading the work of Roger Nelson and his Global Consciousness Project, we again imagined with Pierfrancesco an array of REGs to detect "syncrhonicity" waves—a totally speculative working hypothesis that was just a fancy idea to think about. But then it struck us that we only need two REGs to make sense of direction and correlates of consciousness, we can test for non-locality and causality with just two orchestrated REGs! Of course, any antenna for measuring correlated perturbations in any field has to be a 'dipole', no need for complicated fancy arrays. What a relief. Then we went into relaxation mode as we realised the amount of work this would take and the tests and the availability of travel in these times and ... and ...

A few days ago another bosom friend of mine David Lorimer had told me that a new member, Dr Wolfhardt Janu (later to be known as our very good friend and colleague 'Wolf') had joined the Galileo Commission. They have many interests in common and all about REGs and PEAR and ICRL, David had mentioned, and now I had seen Wolf's message to me earlier that morning, but as I was too busy and entangled in our conversations with Pierfrancesco and Brenda and Pierfrancesco again about our 'REG antennas' I had yet to read it and reply. I left this for after our late dinner and before we retired for the day to our 'binge-watching' as people used to do on these strange days of confinement ...

That was when it hit me! During our conversations about correlations, Wolfhard, our dear Wolf, had sent me a 27-page document, written in an informed and personal, emotional style, about his history of discoveries. He had done what we had speculated we might do in the future. It was as if time had been rewound backwards and forwards.

It had been accomplished. The REG antenna had been invented, implemented and started to produce results. Would you have a look, Wolf kindly asked ... "You bet," I thought, so no binge-watching that night. I read and reread his manuscript. Erudite, technical and to the point, absolutely convincing. Of course, I had been preparing to be convinced all day without knowing it! His experiments from last year (2019) two REGs, as our imaginative ideal antenna wanted, were witnessing and recording amazingly strong and clear correlations when people would gather in a meaningful setting, a ritual of short and raise their state of consciousness as individuals and as a group. Now analysed and tested against control groups and several instances of these ritual gatherings verified, the results were stunning and ground-breaking.

Needless to say, I replied immediately and with equal enthusiasm. In the days that followed, I sent Wolf's manuscript to the friends concerned and invited them all to a meeting. He brought along his, and now our, good friend and colleague Annette Grathoff, a brilliant biologist. And so, from the invisible fertile field of synchronicities that all these wonderful people around us have cultivated and continue to curate with their good intentions and kind attention, our research group, known to its members and friends as "oREGano", sprouted into the light of the sun.

Synchronicities about synchronicity II

Once our small but thriving research group was established, it was as if we had created our own little field of synchronicities. Help and support come from unexpected connections. Friends join in and give their support in wondrously timely ways. As we were talking about this in another virtual GC meeting, another good old friend, Vicente Arraez, joined us. It was as if he was preparing his team of doctors and nurses in the intensive care unit he had been running for years. Immediately, a pilot project was launched, supported by the local university hospital, with the aim of detecting anomalous non-local correlates of consciousness at the time of death. Two other physics professors asked to join the team through GC, and imagine our surprise when we realised that they were Vicente's neighbours in the Alicante area, an hour's drive away. So we now have a Spanish branch of oREGano!

But there was another synchronicity related to the study of synchronicities that played a crucial role in this development. As we were preparing for this project on the correlations of non-local consciousness, with Vicente and Peter Fenwick leading our project, I received word that one of my spiritual mentors and bosom friends, Fotini, had decided it was time to pass on. She was nearly 90 and in frail health. She said goodbye to us and decided to stop eating. The same day I heard the news we had our planned oREGano meeting 'by chance'. I asked Wolf if he could arrange an oREGano observation session for Fotini, who was in her final days. Wolf acted swiftly, hooked up the device and after the meeting Wolf, Annette and I joined our intention and attention for this set-up to accompany Fotini to her passing as she had previously suggested herself. 'I shall give you a sign' she had told me earlier, but at the time

I was unable to grasp what she really meant as I was so emotionally grieving. Now it was clear. And so it was. The REGs responded with an extremely anomalous deviation from random correlations that was undoubtedly a first 'signal' or rather a first sign.

But what kind of sign was that? Fotini was clearly a 'non-dual' or 'enlightened', definitely advanced human being. Although she would abhor such characterisations of herself, which suggests that she was! Does it matter what state and level of consciousness a person is in at the time of death? Or is it that since we all see the Light at that stage, there is no difference? She was not on any medication to alter her mental state. She was lucid and had no fear of death. Are these factors that could influence the characteristics of this 'sign' that comes in synchrony and carries such profound meaning during the ultimate state of consciousness shift at the moment of death? Do tranquillisers or panic and despair prevent or have an effect on such a 'signal' of synchronicity?

But what kind of sign was that? Wolf and Annette were in Vienna, I was in Brussels and Fotini was in Athens. Once again, physical distance was irrelevant to this synchronistic event. It was the noetic, psychic, spiritual-psychological, emotional qualities of our closeness that counted. They were so important. The little ad hoc 'ritual' of connecting the devices made the emotional-noetic distance between all participants even closer, very close, at least in pairs. How can you measure that? But it was synchronous and had an important meaning for all of us. Why is the spatial distance not relevant, but the temporal distance?

Can one wrap ones' mind around all these?

It is about the quality of the connection, not the quantity of the physical distance separating one from another. OK, that's a statement that has some validity, so let's have a look at what else we have here:

The REGs we use utilise inverse diode electron leakage, a purely quantum tunneling phenomenon which is then translated into binary results. That's why external factors such as humidity, temperature, pressure, electromagnetic radiation, etc. can and are shielded. So are we witnessing some kind of entanglement between qualia and quanta in some kind of "quantum consciousness units"? Are these related to those described in Jahn and Dunne's 5M model, or the more recent ones such as d'Ariano-Faggin's scheme and/or Stuart Kaufmann's proposal? Perhaps one could find support in the earlier proposals of Hameroff and Penrose, or even earlier in the schemes of Bohm,

Stapp, or even Wigner and von Neumann? Would these theoretical proposals be assessed in a different way? What indicators can be used to detect this entanglement, and to discriminate between the predictions they make?

Apart from this epistemological exercise, we must also be aware of certain ontological per-epistemological assumptions that shape our ideas of what meaning itself is.

When we discuss causality, causal and a-causal relations, it is important to remember that modern science, as it exists today, observes and seeks to understand only two of the four classical causes. The material cause and the efficient cause. These causes are quantifiable and verifiable by physical means. It leaves out of its discourse the so-called formal and final causes, purpose, intention i.e. teleology. But this is where meaning and semantics reside. It is therefore at least problematic and essentially impossible for modern science to assign meaning from its first principles. This will only be the case if science remains stuck as it is now, oblivious to the qualities of purpose, attention and intention. But this can and will change.

Synchronicity as an observed phenomenon forces us to reconsider the causes we have lost in the process of the scientific enterprise. It re-emphasises wholeness, meaning and purpose. It encourages the study of correlations and correspondences as signatures of underlying interconnectedness and an abundantly rich, multifaceted reality.

My encounter and study of synchronicity reinforces my understanding that we approach reality in the same way that we have retreated from it; or in other words, we approach reality in the same way that reality approaches us. We are part of reality and both reality and ourselves are capable of being known. To be merely understood would be an understatement. Rather, both the reality "out there" and the self "in here" are to be known. Both are participants in gnosis, as the ancients would have it. It is consciousness that makes all this possible in the first place.

Therefore, the meaning of meaning lies in the primordial unfolding of consciousness in the cosmos. Consciousness is the fundamental expression of the immaculate power of love, which is the initiator of its creation. Inevitably, given finite knowledge and understanding, there will always be a gap between what is given and definable and what is real as it comes into existence.

The One, multifaceted and multi-aspect reality that contains and is contained within the whole cosmos, the Unus Mundus (ἕν το πᾶν) of the Ouroboros of Jung and Pauli. It holds the interconnected diversity of its parts by being " within" those parts. When harmonious relationships manifest in this dialectic, they reveal the meaning of meaning.

That's why we must always try to express the inexpressible or speak of the unspeakable. To always keep our minds and systems open. For the cosmos embodies the necessary and sufficient conditions for everything to be as it is, including its logic. It is within such an expanded logic that the apparently paradoxical phenomena of synchronicity find their natural habitat.

References

Janu, W., Basios, V., Moretti, P-F., Merry, P., Grathoff A., & Arráez,V. (2023, June 27). Detecting Deviations from Random Activity as Indications of Consciousness Beyond the Brain.

Linda G. O'Bryant Noetic Sciences Research Research Inaugural Prize Award (2023). Retrieved March 1st 14, 2023, from https://noetic.org/prize-2023/ Full text available onsite.

Basios, V., Moretti, P.-F., Janu, W., & Arráez, V. (2023, May 23-28). The Science of Consciousness 2023 Non-local correlation deviations of randomness as traces of critical consciousness-state changes. Center of Consciousness Studies, University of Arizona at Tucson. Retrieved February 14, 2023, from https://tsc2023-taormina.it/ Primary Topic [04.05] Section TBA; Book of Abstracts pg 136.

Basios, V., Moretti, P.-F., & Janu, W. (2022, April 18-22). The Science of Consciousness 2022 Non-local deviations from randomness occurring during organizational closure. Center for Consciousness Studies, The University of Arizona at Tucson. Retrieved February 14, 2023, from https://consciousness.arizona.edu/2022-conference Primary Topic [04.05] Section C5; Book of Abstracts pg 136.

Basios, V., & Moretti, P.-F. (2020, June 8). Experiencias Cercanas a La Muerte (ECM). Fundación Metta Hospice. Retrieved February 14, 2023, from https://fundacionmetta.org/articulo.php?articulo=50&tema=Actualidad
Note: English version also available on the site

Bohm, D., & Hiley, B. J. (1995). *The undivided universe: an ontological interpretation of quantum theory.* Routledge.

D'Ariano, G. M. (2017). Physics Without Physics: The Power of Information-theoretical Principles. International Journal of Theoretical Physics, 56, 97–128. https://doi.org/10.1007/s10773-016-3172-y

D'Ariano, G.M., Faggin, F. (2022). Hard Problem and Free Will: An Information-Theoretical Approach. In Scardigli, F. (eds) Artificial Intelligence Versus Natural Intelligence (pp. 145-192). Springer, Cham. https://doi.org/10.1007/978-3-030-85480-5

Fuchs, C. A., & Khrennikov, A. (2021). Preface to Special Issue: Quantum Information Revolution: Impact to Foundations. Foundations of Physics, 50, 1757–1761, 1757–1761. https://doi.org/https://doi.org/10.1007/s10701-020-00401-0

Jahn, R. G., & Dunne, B. J. (1986). On the quantum mechanics of consciousness, with application to anomalous phenomena. Foundations of Physics, 16(8), 721–772. https://doi.org/10.1007/bf00735378

Jahn, R. G., & Dunne, B. J. (2005). Consciousness, Information, and Living Systems. Cellular & Molecular Biology, 51, 703–714. https://doi.org/10.1170/T679

Kauffman, S. A., & Radin, D. (2023). Quantum aspects of the brain-mind relationship: A hypothesis with supporting evidence. Biosystems, 223, 104820 https://doi.org/10.1016/j.biosystems.2022.104820

Nadeau, R., & Kafatos, M. (2001). The Non-Local Universe: The New Physics and Matters of the Mind. Oxford University Press.

Radin, D. I., & Nelson, R. D. (1989). Evidence for consciousness-related anomalies in random physical systems. Foundations of Physics, 19(12), 1499–1514. https://doi.org/10.1007/bf00732509

Wahbeh, H., Radin, D., Cannard, C., & Delorme, A. (2022). What if consciousness is not an emergent property of the brain? observational and empirical challenges to materialistic models. Frontiers in Psychology, 13. https://doi.org/10.3389/fpsyg.2022.955594

Main R. *Breaking the Spell of Disenchantment: Mystery, Meaning and Metaphysics in the Work of C.G. Jung*. Zurich Lecture Series in Analytic Psychology, Vol. 8, Chiron Publications, 2022.

Beitman B. *Meaningful Coincidences: How and why Synchronicity and Serendipity Happen*. Park Street Press, 2022.

Tarnas, R. *Cosmos and Psyche: Intimations of a New World View*. Viking, 2006.

Schwartz, GE. *Super Synchronicity: Where Science and Spirit Meet*. Param Media, 2017.

Chapter Sixteen

Deus Ex-Machina

BERNARD BEITMAN, MD

Bernard Beitman, MD is the first psychiatrist since Carl Jung to organize the field of meaningful coincidences. He is the author or editor of seven psychiatry books, recipient of two national psychiatry awards for his psychotherapy training program, and former chair of psychiatry at the University of Missouri-Columbia. He is the Founding Director and President of The Coincidence Project. His first coincidence book, *Connecting with Coincidence* (2016) was followed by *Meaningful Coincidences: How and Why Synchronicity and Serendipity Happen* (2022). His coincidence memoir, *The Adventures of Dr. Coincidence: A Psychiatrist's life of Synchronicity and Serendipity* is in press.

Anecdotal evidence from creating The Coincidence Project (TCP) suggests that immersion in the study and promotion of meaningful coincidences, increases the likelihood that meaningful coincidences will occur as part of this creative process. We have collected numerous such incidents. In this paper, I describe three crucial synchronicities that appeared "out of the blue" just when what TCP needed was offered. While many synchronicities can be described as precognition and clairvoyance as did Jung (1973) in his classic monograph or partially explained by personal agency (Beitman, 2022), the need fulfilling timeliness of the synchronicities reported here seem to emerge from a higher level of mystery.

Unexpected money

In November, 2022, soon after the incorporation on April 14, 2022 of The Coincidence Project as a non-profit in the state of Virginia, I received a check in the mail for $1000 for TCP "to be used in any way you see fit". Our fundraising consultant gave me the donor's contact information. At the end of a 30 minute conversation, she decided to send us a check for $25,000 which was utterly necessary to get us through the first year.

One editor quits, a better one arrives

I recruited an old friend to edit the stories sent to the TCP website (https://thecoincidenceproject.net/) for publication. She helped to establish the necessary guidelines and then decided this work was not for her. Two days later on March 31, 2023 I received an email from Ken Bell: *"I've now worked my way through all the Coincidence Project podcast episodes, and the more I listen, the more I feel connected with Dr. Beitman's work and its importance.*

Which brings me to my second point: I want to help! I'm not sure how I could become involved as a volunteer, but I'm feeling a pull that I should offer to help this movement gain steam.

I have a 30-year background in marketing and communications, so perhaps I could help with web content, emails, or other audience communications. In addition to being a writer, I have a background in graphic design, so could potentially help with branding or visual content. I also have a great deal of experience promoting and running live webinars if that would be useful (I'm hosting a webinar next week with 5,000 registrants)."

And he is a good online editor! Just what we needed and more.

Crawling out on a limb

In March 2023, I offered to do a webinar on Medical Synchronicities for the alumni of Andrew Weil's Integrative Medicine Fellowship at the University of Arizona. The staff were interested and so I was out on a limb. I had plenty of psychiatric and psychotherapy examples. I needed cases from non-psychiatrists. On April 2, 2023 James Schemmel emailed me this out of the blue: *"I'm a 62 year old otolaryngologist. Multiple very interesting Coincidences in my life. Many seem to involve the practice of medicine. What I find most interesting is their impact on my interaction with different patients."*

I requested case examples and he delivered beauties! Two cases were examples of precognition. Here is one of them:

On a Sunday morning while reading a book. I had some sort of vision or dream. It came into my mind of a patient I had never met before. The patient was in the hospital at another service. He just happen to have a tracheostomy. The vision or dream that came to me was the patient bleeding from his tracheostomy site. With a host of hospital personnel surrounding his bed trying to sort out the situation. That evening around approximately 10 o'clock I was called to the hospital to see a patient. When I walked into the room, it was the exact same scene that I had witnessed earlier that morning, the same people standing around the patient.

It was very peaceful and surreal. I felt no stress or anxiety. Things went extremely smoothly because I had already anticipated the situation.

Explanations

The editor Ken Bell was seeking to volunteer for TCP and the physician, James Schemmel, was seeking dialogue with me. These instances exemplify the idea that "what you are seeking is also seeking you." Together they confirm this basic synchronicity principle. Then how do we explain this principle?

A thorough description often becomes a good explanation. The intention behind and timing of each contributor's decision to contact me are keys to a full description. The donor wrote "This sounds like an exciting project, and is the basic premise of my own life. I was lying in bed on forced bed rest for too long. I read an article about your project and wanted to raise my hand and say this is the way the world works."

The editor Ken Bell read about my book in a front page article in the Los Angeles Times, the paper edition of which appeared in December 6, 2022. He contacted me about four months later, a few days after our previous website editor quit. He wrote, "I'm feeling a pull that I should offer to help this movement gain steam." This pull was precisely timed to match our current need. The physician was curious about his experiences of meaningful coincidences and apparently wanted more information. He did not report the reason behind the timing of his decision to contact me.

Using Ken Bell's explicit comment of "feeling a pull", it is possible that the donor and the physician also felt a pull or an urge to contact me. Following an intuitive urge to act is commonly reported by high frequency coinciders, those who experience many synchronicities. If so, then the next question becomes what creates the pull/urge?

We are all connected

Indigenous people and many modern religions tell us that we are all connected. Science has accumulated sufficient evidence to support this conclusion (Currivan, 2022). To fully know our connections with each other, plants, animals, and the earth itself, requires experiences of these connections. Cognitive knowledge is insufficient. The emotional surge of a connectedness experiences strengthens this knowledge. Meaningful coincidences of synchronicity and serendipity are usually accompanied by varying degrees of emotional surges the strongest of which Jung called numinous, a momentary experience of oneness.

The emotions accompanying these experiences also illuminate the invisible currents that connect us with each other; they show us how we are connected. One common example is simulpathity—experiencing the distress of a loved one at a distance. This experience was at one time called telepathy.

The three stories described here illustrate the further reaches of knowing the distress of someone at a distance. None of these three people knew me and yet each decided to help me in my time of need. How did they register my needs as the founder of TCP?

We can conclude that under the right conditions, that which you are seeking is also seeking you. And that by following some kind of hunch, intuition, urge or pull, mutual benefit can bloom. What are those conditions? Being immersed in the experience of the study and promotion of synchronicity appears to be important. Personal agency, the willingness to follow inner guidance appears to be another. The most improbable aspect of these stories is the timing. In synchronicity timing is often the source of the emotional surge as in being in the right place at the right time, a commonly reported synchronicity. As suggested by mystics and quantum physics, time may be a fundamental property of living on Earth but not of the hidden realities indicated by many synchronicities.

Through simulpathity, these three people may have subconsciously felt my distress and converted their experience of my distress into an urge to help me.

Mysticism and science converge

Although many human beings deny the reality of global warming, our planet is heating up. We are like the proverbial frog swimming in a pot of slowly heated water, barely noticing the increasing temperature. Many people are noticing, particularly younger people who are inheriting a diseased biosphere from older people and most acutely by subsistence level farmers all over the world who had relied on regular weather patterns nourish their crops.

As science increasingly supports our interrelationships and mystics continue to tell us we are all One, humanity must become practical. We must band together with the help of synchronicity to see the far reaches of our interconnectedness as these three stories begin to illustrate. We not only feel the needs of loved ones, as is commonly reported by identical twins, parents and children and close friends (Beitman, 2022) but also with people we do not know.

Drawn to an existing future?

John Townley is currently working on a book-to-film project about Austrian biologist Paul Kammerer and his seminal 1919 *Das Gesetz der Serie* which introduced the structural concept of seriality, tapped by Jung in his speculation on synchronicity. I asked him for an explanation: *I think that a lot of "feeling the pull" or 'hunches" or little beckoning event-windows are not coming from you per se but to you from the larger involvement that is already a part of in the future, where the gravity of what you are involved in (with TCP here but it could be anyone's situation) reaches back and tugs you forward. You didn't just need a new editor—one was already interested but didn't know it quite until the need arose and he, and you, got the call from be the larger/"later" TCP reality and then it happened, on time. Same with the medical story. It is part of both the "attraction hypothesis" of seriality and the greater persistence/gravity/inertia of the later and larger entity of which you are both part and whose tendrils knocked on your window. When the need or direction gets stronger, so does the signal and ever more appropriate people can pick it up. Individually, we have spotty receivers and scientifically we don't know the frequencies to explore yet. AI and lots of data (such as TCP is collecting) may help with that.*

Future research

I make two research proposals: 1) to explicate the range of patterns in synchronicity stories using artificial intelligence and 2) to develop a conceptual model of our mental atmosphere, the psychosphere.

Patterns

A new science begins with description and categorization. I propose that there are at least four kinds of meaningful coincidences: synchronicity, serendipity, seriality and simulpathity. In his seminal treatise on synchronicity, Jung used the term synchronicity primarily to explain meaningful coincidences. His description of the sequential appearance of fish fits within seriality. In *Memory Dreams and Reflections,* his recounting his severe headache while in a hotel room correlated in time with the suicide by bullet to the head of one of his patients. This coincidence would fit within simulpathity. These four types share two basic elements: mental events and objective events or mind and object. Mind and object similarities arise in many different ways that could be studied by using Normal Language Processing (NLP) to search out patterns that human minds could miss.

Once the patterns of meaningful coincidences are clarified, I propose that we develop an app that links people who experience similar types of meaningful coincidences. These patterns can include: being in the right place at the right time; remarkable series of similar events like repeated 11:11, developing intense relationships in the present with previously unknown people from similar geographically based backgrounds; repeated coincidences between paintings and photographs and many, many others. These linkages can accelerate recognition of our interconnectedness and the usefulness of synchronicity in illuminating the invisible currents connecting and uniting us.

Psychosphere

I propose the existence of our mental atmosphere, the psychosphere, that surrounds us like our air atmosphere and in which all beings are immersed. Within the flux of the psychosphere, varieties of energy and information move the way gases, clouds, wind, water vapors, particles, and electromagnetic radiation move in our air atmosphere. The energy information of the psychosphere includes the full spectrum of human ideas and feelings and behaviors. Just as our living bodies breathe in oxygen and breathe out carbon dioxide, each human mind interchanges ideas and energy and behavior patterns with the psychosphere by expressing them and drawing them in.

The French philosopher and Jesuit priest Teilhard de Chardin and the biogeochemist Vladimir Vernadsky called it the noosphere, which means "the sphere of mind." I coined the new term, believing it to be more accessible. The psychosphere emerges from the biosphere. The biosphere is made up of the parts of the Earth where life exists. The biosphere extends from the deepest root systems of trees, to the dark environment of ocean trenches, to lush rain forests, and high mountaintops. The solid surface layer of the Earth is the lithosphere. The atmosphere is the layer of air that stretches above the lithosphere. The Earth's water—on the surface, in the ground, and in the air—makes up the hydrosphere. The psychosphere is a part of nature just as are the lithosphere, hydrosphere, atmosphere, and biosphere. The psychosphere emerges through, and is constituted by, the interaction of human minds as well as the consciousness of plants and animals and the earth itself. (Beitman, 2022).

I propose that our interconnectedness takes place through the psychosphere and that a model for telepathy and simultaneous independent discovery can be developed to research the workings of the psychosphere.

References

Jung, C.G. (1973). *Synchronicity.* Princeton Press.

Jung, C.G. (1963). *Memories, Dreams and Reflections.* Princeton Press.

Beitman, B.D. (2022). *Meaningful Coincidences.* Inner Traditions.

Currivan, J. (2022). *The Story of Gaia.* Inner Traditions.

Chapter Seventeen

My Synchronistic Journey

SOPHIA DEMAS, MED

Sophia Demas has enjoyed three diverse careers: a decade in architecture that included working with the visionary Dr. R. Buckminster Fuller, ten years running her own couture fashion design business, and after returning to school to earn a master's degree in counseling psychology at age fifty, became a mental health therapist. She created *Living a Fearless Life,* a 12-workshop program designed to enhance self-esteem in at-risk women. Currently, she serves the Salvation Army as an in-house counselor in a residential facility for ex-trafficked women. Sophia lives in Philadelphia with her husband Frank, an architect, and is sublimely happy.

Coincidences, which are chance occurrences and synchronicities, meaningful coincidences, are commonplace. In fact, many people have experienced life-changing synchronicities such as having an unforeseen obstacle delay your vacation, only to find out that the destination was flattened by a devastating tornado or, for no reason, you walked a different route to work and when you decided to stop for a coffee you ended up meeting your future spouse. What I have struggled to understand, however, is why I, an ordinary person, have experienced an inordinate number of synchronicities that have literally molded my life, and what could have allowed me to be open to them.

I am convinced that the seeds of my receptivity were sown in early childhood. A little background: My father had immigrated from Greece in 1909 by himself when he was just fifteen. He worked multiple jobs around the country, finally settling on the assembly line at the Ford Motor Company in Dearborn, Michigan in 1922. It wasn't until forty-two years later that he returned to the "old country," married my mother fourteen days after meeting her, and brought her back to the States where I was born a year later. We had no relatives here. When I was seven, my father retired and we moved to Greece, where instantly I had a boisterous extended family. It was magical.

However, Greece's geographic vulnerability to the USSR during the Cuban missile crisis scared him into moving us back to the States. We ended up in Portland, Oregon, where my father bought a house near the Greek Orthodox church which, aside from her family, became the center of my mother's life. She had indoctrinated me from birth, using the Greek diminutive, she inferred that "little" Jesus, who loved children, and the "little" Virgin Mary were my friends whom I could ask for anything I wanted, and who would always protect me. Consciously beseeching these unseen beings seemed perfectly natural to me.

It wasn't until I was nineteen that I began to recognize that something larger than me was pulling the strings. I was agonizing over two issues that seemed to be unsolvable—I obsessed about what people thought of me and I had bought into my mother's law of no sex until marriage that now wasn't working very well for me. At the time, I looked upon marriage as unsavory and did not want to die a virgin. A timely event involving a man I had met whom I was extremely attracted to and a momentous dream resulted in not only the resolution of my emotional distress, but also transformed me as a person. Of course, he was the one with whom I had my first sexual experience and since then I have never cared about what anyone thought of me again. I had palpably entered womanhood. The awakening made me start paying attention.

I found that the more elevated my awareness, the sharper my intuition became and the more I trusted it for guidance. This cultivated inner voice has enabled me to recognize coincidences as divine communications that result in miracles—receiving exactly what I need at just the right time.

These experiences have molded my ontological understanding of being. It is based on the premise that there is a Divine Intelligence/God that permeates everything and transcends time and space with whom we can have a personal relationship. This Higher Power promotes the reality of Oneness/Love, as opposed to the ego whose mission is to make us believe that we are separate from one another and to keep us divided. Universal Consciousness wants us to be happy and find our true purpose. It is within these realms of consciousness that God, who knows what's better for us than we do, communicates with us in many clever ways via our inner voice, a stranger's message, dreams, and by putting coincidences in our path to nudge us along. Unfortunately, we were given free will that results in us resisting or ignoring these signs. By taking action on a coincidence we have the opportunity to make it meaningful by becoming co-creators with the Divine in manifesting the fulfillment of our needs and desires. In turn, my synchronistic experiences validate my ontology.

To focus on my career, I had decided not to get married or have children. The last thing I wanted to do was to write a book. The Universe, however, had other plans. Thank God I listened to those nudgings! I had avoided marriage until I was forty-five when I married Frank. My friend and mentor, an Albanian Orthodox priest, who had launched shelters for homeless teens in NJ, then introduced me to a beautiful teenage girl from Turkey who was living in one of them. Frank and I adopted her and Mother's Day became my favorite holiday. And, besides enjoying three successful careers, I've written two books!

Although I had been raised a Greek Orthodox Christian, I have been greatly influenced by the teachings of the Toltecs, Buddha, and especially Jesus, not only by his words, but by his actions. I also studied the Torah with an Orthodox Jewish rabbi. The heart of these traditions does not differ much from one to another. It was *A Course in Miracles* (ACIM), however, that brought me to a deep understanding of the "Christ Mind," the belief that you, me, God, and consciousness, are All One. Principles of quantum mechanics have helped me better understand this oneness.

According to ACIM, there is no difference between a big or small miracle. I feel the same way about synchronicities, though some have played a more pivotal role in my life than others. The following two improbable synchronicities contributed significantly to the festivities of my wedding. Not only did they save us a good deal of money, they saved us time since we only had three months to plan the event:

1) I dialed my friend Ann at her studio, and a man answered. When I asked to whom I was speaking, the man said, "Sophia? It's Jerry. You called me! I'm late for a meeting. Can I call you back this afternoon?" I made him promise. Jerry was someone I knew socially, who had recently purchased the historic Barclay Hotel in Philadelphia. He did return the call and I found out that I had misdialed the last digit of Ann's number. As a result, of all the people in Philadelphia, I reached an acquaintance on the phone who recognized my voice, offered me the hotel's grand ballroom for our wedding reception as a wedding present, and threw in a two-night stay in the presidential suite. I merrily granted his request that in return, he and his girlfriend be invited to the wedding.

2) It was winter and a woman who identified herself as Lisamarie called me at home to tell me that she had found my driver's license in the snow. She lived only a few blocks away and wanted to know how she could return it to me. I thanked her profusely and asked her to drop it through my mail slot. Five months later, two months before the wedding, I received another call from Lisamarie. "I know this will sound strange," she said haltingly, "but I just found your ATM card." I was stunned. The ATM machine I often used spat out the card before completing a transaction. This machine kept the card until one answered the question, "Would you like to make another transaction?" Forgetting the card, I left without answering. The next person could have cleaned out my bank account! Lucky for me, the next person was Lisamarie. Now she wanted to know if she should slide the card through my mail slot. "No! We have to meet!" I exclaimed incredulously and invited her to my studio. A lovely young woman arrived and I invited her in for coffee and cookies. How could I thank her enough? She revealed that she was a concert violinist. I told her of our upcoming wedding and how we were trying to decide on the church music. I asked her if she had ever played "Concierto de' Aranjuez" by Joaquín Rodrigo for the classical guitar. Of course she had. "I would love to play it for you as my wedding present," she blurted out. "I'll get my friend Alan to accompany me on the classical guitar, and you can pay him something."

What was the probability that the same woman who found my driver's license would also find another one of my cards? At just the right time! This resulted in me walking down the aisle to one of my favorite adagios.

Although the Universe cleverly handed me two meaningful gifts on a silver platter, let's not forget the hand I played in both synchronicities. It would have been perfectly normal to have replied to Jerry's offer to call me back with, "Oh, don't bother—you were a wrong number," since I really had nothing to say to him. Or I could have simply thanked Lisamarie for finding my ATM card and have her slide it through the mail slot as she had done with my driver's license. My intuition, however, knew that these occurrences were bigger than me. This propelled me to take action, demanding to have further communication with both people involved. By doing so I stepped up to Divine Intelligence and the synchronicities were co-created, resulting in exactly what I needed at just the right time.

A string of three events in the following story and what it yielded is no less improbable. Although I was thrilled and grateful for the timely and meaningful synchronicities that made our wedding even more memorable, I place more importance on this one because it saved my marriage from divorce.

Six months into my relationship with Frank, I discovered that he had major anger issues. As a result of unresolved negative childhood experiences, he would perceive an innocent gesture as demeaning and instead of qualifying it, this wonderful, mild-mannered man would fly into a fiery rage. Regardless, I married him because I saw how much he wanted the relationship and the real effort he made to heal himself. I threatened to leave twice. After the first time he found an excellent therapist and the second, he enrolled himself in Menergy, a men's group for ragers. However, the explosive episodes continued and finally I had had enough. I left and moved in with a friend. This only fed his anger and precluded any negotiating. I made up my mind—I never wanted to lay eyes on this jerk ever again. The following three synchronicities happened within two months:

1) With the divorce attorney's card in hand, I excitedly opened my well-worn ACIM to a random page, hoping to glean words of wisdom and guidance for my new-found freedom. To my dismay, the chapter was titled, "The Healed Relationship." My rule was to read whatever I opened the book to, regardless of how much I didn't want to. I understood the chapter's message to mean that God is in everyone, and if your brother comes to you with his hand outstretched seeking connection and you say no to him, you are saying no to God. I decided that if Frank wanted to communicate, it wouldn't hurt to say yes, but it wouldn't change my decision one bit.

2) Frank called the very next day asking to meet for a talk. I agreed and invited him to the home where I was staying. I stood and told him that the only reason he was sitting across the table from me was because of a miracle. "I have a miracle for you, too," he replied. I said, "You go first." Words like this were not part of Frank's lexicon. As he removed his wedding ring, he told me that while taking apart the fountain in our garden, the heavy cement cylindrical part fell on his hand. I shot a glance at it. If that was true, his hand should be hamburger, but all I saw was a bruise with a scratch. "My wedding ring saved it," he continued, holding up the squashed band. This completely disarmed me and we began talking. The love was solidly there and we started seeing each other again. After six months apart I moved back home.

3) I was planning to go to Maui, Hawaii, for my best friend's wedding to be her matron-of-honor for the third time. I wanted Frank to come with me, but there was trepidation. Although he had sabotaged many romantic

getaways in the past, I decided to go all in and trust him. In a rush to take a cab to the airport, I mistakenly grabbed *Endless Energy* instead of a novel. Written by Debra Greene, a kinesiologist, it was about the healing aspects of energy work, not your optimal beach read. Reading it on the plane, I was fascinated with the consciousness-based modality she used in her sessions. As I was flipping to her sources in the back of the book, I landed on a page with her contact information. I was stunned. Not only did she live on Maui, but her office was on the same road where we had rented a condo. I showed it Frank. Needless to say, we each had a healing session with Debra. That was in 2009. Frank hasn't had one anger episode since. By the grace of God, I am married to my dream companion and co-pilot on this life's journey!

What my mother instilled in me since infancy is deeply rooted. I ask for what I want and express gratitude all day long. Via consciousness, I have developed a personal relationship with God that I rely on for the guidance I receive through my intuition. In my book, *The Divine Language of Coincidence,* I have retranslated my go-to-prayer, the Lord's Prayer, from Greek to English to restore the nuances lost in translation. It received the blessings of my friend, Fr. Dr. Demetrios Constantelos, a renowned theologian and Greek scholar. In the first line of the original Greek version, *Heavens* is plural to encompass all realms. The first line of my translation also addresses these realms:

"Our Creator who resides in sound, light, and vibration."

This confirms for me that Universal Consciousness permeates all realms and that synchronicities can only occur in an intelligent universe.

Chapter Eighteen

When the Everyday World Comes Apart

IMANTS BARUŠS, PHD

Imants Baruŝs, PhD is a professor of psychology at King's University College at Western University where he teaches courses about consciousness, altered states of consciousness, and the psychology of religion and spirituality. He is the author of numerous papers and books including *Radical Transformation* and *Death as An Altered State of Consciousness.*

A synchronicity occurred for me which loosened my conceptual grip on the everyday world and gave me reason to believe that reality is more interesting than we assume it to be. Embedded in the synchronicity was an additional striking anomaly. Synchronicity refers to a constellation of events that is typically identified by four characteristics: occurrence within a limited time span, stochastic improbability, connection through the meanings that we attribute to the relationships of the individual events, and not apparently connected through cause-effect chains of local interactions (Butzer, 2021). The expression "anomalous phenomena" designates the occurrence of events that violate conventional interpretations of reality. In my case, the anomalous event was anomalous object displacement whereby an object changes location, usually while unobserved, without any viable means for achieving that displacement.

My experience

I sometimes do distance healing when a person asks for my help. On November 13, 2019, through an intermediary, I had agreed to help someone, let me call them Alex, who was dying of cancerC someone I have never met. I sat down at my desk in my home office and began. I usually base what I do on a set of techniques called ME, or Matrix Energetics, created by Richard Bartlett (Bartlett, 2007). I have described my experiences with ME elsewhere (Baruŝs,

2013). All that is necessary to know here is that I work with visualization. In this case, I was imagining the removal of the cancer. And because Alex was in such late stages of the disease, it seemed to me that that removal would not be through the stimulation of a biological healing process, but an anomalous disconnection of the cancer from their body.

The following day I was getting ready to drive to the college where I teach. I have two sets of keys—one for my personal life and one for school. The school keys are on a tight key ring and include door keys, filing cabinet keys, and a mailbox key for a total of seven or eight keys. I put those in my pocket and used the personal set to open and drive my car. As I was driving, I was thinking about the previous evening's distance healing session and my effort to remove the cancer from Alex's body. I got to school in about fifteen minutes, parked my car, and ascended the three flights of stairs to my office.

Standing in front of my office door, I reached into my pocket and pulled out the school keys. I was looking for the key for my door. I had three door keys that all looked pretty much alike. However, I knew which was the key for my office because it was right beside a filing cabinet key that had a black, plastic top. That filing cabinet key was the only one with a plastic top, and it was black, so it was easy to recognize. After fumbling with the keys for a few moments, I realized that the filing cabinet key with the black top was missing from the key ring. That is why I could not identify the office key. I reached into my pocket, and there it was. The black key. All by itself. No longer on the key ring. The key ring is tight, and requires twisting and pushing a key with considerable force using both hands in order to remove it from the ring. The keys cannot just "fall off." They had never previously "fallen off" nor have they "fallen off" since. Nor is it reasonable to suppose that they could just "fall off." I inspected the black key carefully because I assumed that the plastic top must have broken off, releasing the key from the key ring. But there was nothing wrong with the top of the key. It was intact. I went into my office, prepared for my class, and went across the campus to teach my students.

The topic for my class that day was anomalous mind-matter interactions. The assigned reading was Chapter 6, "Direct Mental Influence," from *Transcendent Mind* (Barušs & Mossbridge, 2017). I considered telling my students about my experience with the black key, but decided against it, in spite of the relevance to that week's topic. I did not want to risk having some of them complain to the academic dean that I had become senile and should be dismissed, as had occurred on a previous occasion when I had shared my personal impressions. I did tell a few of the students who were lingering after class and who were particularly interested in the course content.

The personal meaning of my experience

I realized right away that this was an anomalous occurrence. Anomalous phenomena usually show up for me with considerable room for doubt as to their anomalous nature. Precognitive dreams, remote viewing a problem at the back of a person's neck, seeing an imaginary black cat rubbing itself against my leg, a black butterfly landing on my knee, even anomalous object displacement, whereby a long-lost necklace shows up in the middle of a corridor in my house. These are ambiguous situations with considerable latitude for interpretation. What happened to me that afternoon left little wiggle room. I had started out with keys nicely fastened to a key ring and ended up fifteen minutes later with one of them disconnected from that key ring. All of this during an otherwise boring, normal day. This was a blatant violation of Newton's laws of motion. There was no escape from that observation.

The fact that it was the black key that had disconnected, instead of one of the other keys, had symbolic significance for me. I considered two meanings. The first was that the black key represented cancer that had disconnected from Alex's body. The second, that the black key represented Alex, disconnecting themselves from the world of the living, represented by the ring with the rest of the keys; in other words, death. It was also meaningful to me that the disconnection of cancer from the body, and curiosity about Alex's health, had been on my mind as I had been driving, and that my thoughts were apparently instantiated symbolically with the removal of the key. The coincidences of the previous evening's distance healing, the disconnection of the key as I was still thinking about the healing, and the class lecture about anomalous mind-body interactions, together create a synchronicity.

For all that I have had unusual experiences and have been teaching university students about such phenomena for forty years, I still question their occurrence as anomalies. Perhaps there is something that I have overlooked. Perhaps I have made a mistake. Perhaps life really is brutal and meaningless. You get cancer. You die. Alex died. Not nice. That is life. Regression to the mean. But I took notice of this event. In the middle of everyday life with its apparent indifferent oppression, a key comes apart from a key ring in my pocket. Anomalous object displacement. And the mundane world is no longer mundane. We do not already know everything after all. An opening is created into something interesting. Promising. Hopeful. Magical. Playful. Reality can be like that. In a moment, transformed. Can I learn to harness whatever this was that happened? As a healer, could this open up a bag of tools available to me? What else is possible? How outrageous is nature, really? Such questioning presents an additional symbolic interpretation of the anomalous object displacement, namely, that the set of keys coming apart represents the coming apart of conventional ways of thinking.

The Explanation

Academics, such as myself, are saturated with a materialist world view to such an extent that, even when we intellectually reject materialist ideology, there are nonetheless implicit cognitive schemata that we continue using for making sense of reality. One of those is the notion that the universe is mechanical. That it is made out of insentient bits of matter that bump into one another with predictable trajectories. The universe is meaningless. Humans are capable of meaning-making, but only humans. We are the sole beings for whom anything means anything. Life is meaningless unless we ascribe meaning of some sort to it. We are the light in a dark room. But what if that assumption, clearly a vestige of materialism, is wrong? What if meaning is inherent in nature and our meaning-making is simply one aspect of the meaning-making of the universe? What if nature can understand things in the same ways that humans can? What if reality is inherently suffused with meaning?

To incorporate such ideas, I generated a theory of meaning fields. Elsewhere, I have written an overview of the theory (Barušs, 2018), a detailed exposition (Barušs, 2021), and developed a mathematical model (Barušs, 2019). The thrust of the theory is that reality is structured along lines of meaning rather than physical proximity and physical influence. The core concept is that the everyday world that we experience, including time and space, is not a continuous stand-alone reality, but a flickering moment-by-moment projection from a deep, timeless and spaceless transcendent consciousness that contains everything that is possible all at once. At any given moment, anything could emerge from that deep consciousness. However, continuity of expression is provided by meaning fields that structure the explication. These meaning fields can be thought of as entities that understand connotative and denotative meaning and have remote viewing and remote influencing capabilities. They can interact with humans, and humans have some ability to choose which meaning fields structure their experience.

Applying this theory to my experience, I am usually living under the meaning field of ordinary reality in which keys stay on their key rings. However, through engaging in ME, in an explicit attempt to step out of ordinary reality, I tuned to a meaning field that allowed for anomalous object displacement and my key went from being in one location, on the key ring, to a different location, not on the key ring. How could it do that? Because there are no keys or key rings. Those are all momentary projections from a deep consciousness which can explicate different events at different moments in ordinary time. It is just a matter of learning how to tune between meaning fields. Because humans can influence meaning fields, we could experience modified meaning fields that align with our nonconscious expectations of what is possible. So we might only get some ambiguous anomalous phenomena once in a while or, if we have opened up, perhaps awesome striking events that shake us to the core of our being. And the more we get shaken, the more we open ourselves to the possibility of anomalous phenomena and move along a more expanded path. And with that, there is an additional interpretation for the fact that it was a key that came apart from a key ring. I am continuously seeking the key to the mysteries of the universe and here it is, literally a key. And it is a black key signifying the black sheep that strays from the flock immersed in conventional thinking. The key to the mysteries is the coming apart of ordinary reality to open us to deeper meaning. Mine was a rich experience involving anomalous object displacement and synchronicity that has given me a valuable foothold for my process of self-transformation.

References

Bartlett, R. (2007). *Matrix energetics: The science and art of transformation.* New York: Atria Books.

Barušs, I. (2013). *The impossible happens: A scientist's personal discovery of the extraordinary nature of reality.* Alresford, Hampshire, UK: John Hunt Publishing.

Barušs, I. (2018). Meaning fields: Meaning beyond the human as a resolution of boundary problems introduced by nonlocality. *EdgeScience, 35,* 8–11.

Barušs, I. (2019). Categorical modelling of conscious states. *Consciousness: Ideas and Research for the Twenty-First Century, 7*(7), 1–10.

Barušs, I. (2021). *Radical transformation: The unexpected interplay of consciousness and reality.* Exeter, UK: Imprint Academic.

Barušs, I. & Mossbridge, J. (2017). *Transcendent mind: Rethinking the science of consciousness.* Washington, DC: American Psychological Association.

Butzer, B. (2021). Does synchronicity point us towards the fundamental nature of consciousness? An exploration of psychology, ontology, and research prospects. *Journal of Consciousness Studies, 28*(3–4), 29–54.

Chapter Nineteen

The Case of Gail Windwalker: Spontaneous Mediumship as Synchronicity

MICHAEL A. KRIEGER, MD
AND SUNSHINE L. KESSLER

Michael A. Krieger, MD majored in physics and psychology at Rutgers College and earned his MD at GWU Medical School. He completed internal medicine internship at the GWU Hospital and psychiatry residency at Boston University Medical Center. He is board certified in Adult, Addictions, and Geriatric Psychiatry. He is a graduate of the Boston Psychoanalytic Society and Institute. He was staff psychiatrist at the outpatient mental health clinic of the VA Hospital in Bedford, Massachusetts, where he taught medical students, served as medical staff president, and chaired the Ethics Advisory Committee. Now semi-retired, he maintains a solo private practice from his home

Sunshine L. Kessler is a digital nomad with an academic background in the study of consciousness. A craft beer enthusiast and international beer judge, Sunshine has somehow managed to visit all 7 continents and over 100 countries on a writer's salary. She most appreciates visits to remote or challenging places like North Korea, where she was among the first group of foreigners to tour local breweries. An avowed psychonaut, Sunshine started the Lucid Humanity initiative for understanding consciousness and the nature of reality. Ashtanga yoga and meditation practices keep her feet on the ground.

Introduction: The flip

Upon graduation from residency in 1993, I took a position as staff psychiatrist at the VA in Bedford, a suburb of Boston. Thus began what would become my 29-year career in their outpatient mental health clinic. In 2005 I began my psychoanalytic training. It was through this training I learned how to reside in a world of ambiguity, maintaining an open "play space," where patient and analyst creatively co-construct a new reality while keeping an open mind to the infinite possibilities. However, it would be more than another decade before I would learn just how closed my mind had been.

I had been teaching medical students that belief in the occult or preoccupation with synchronicities is consistent with the "odd beliefs," "ideas of reference," "magical thinking," and "unusual perceptual experiences" characteristic of Schizotypal Personality Disorder. Only later did I learn that while psi experiences can be symptomatic of some psychopathologies, an abundance of research has revealed "the potential for spiritual experiences to treat disorders such as addiction, depression and anxiety," (Corneille & Luke, 2021).

Then, something happened to turn my worldview upside down. Like the paradigm shifts Dr. Jeffrey J. Kripal recounts in his book *The Flip: Epiphanies of Mind and the Future of Knowledge*, I would experience my own ontological flip. In retrospect, I had overlooked many events preceding this one that could have altered my worldview. My flip would require a bigger threshold event. This event occurred on May 26, 2016. On that day, I had an experience that I have since learned is called "spontaneous mediumship."

What is spontaneous mediumship?

Spontaneous mediumship experiences (SMEs) are considered a subset of after-death communication (ADC), in which "a living person senses the presence of a discarnate (deceased person) who may or may not overtly communicate a message," (Holden, Foster, et al., 2014, p. 68–69). In other words, "I see dead people." Mediumship is a more specific sub-phenomenon of ADC in which, "instead of the ADCr directly perceiving the discarnate, a third person acts as a medium of communication between the discarnate and the living person... the message is not for the medium but for the sitter," (Holden, Foster, et al., 2014, p. 70). As with Victorian seances, the majority of mediumship experiences are intentional, not spontaneous (Foster, Kahoe, et al., 2018). Because they are relatively rare compared to other ADCs, SMEs are in the "infancy of scientific investigation," (Foster, Lee, et al., 2015, p. 45). I hope my case study adds to the growing body of literature on the epistemic value of SMEs for understanding consciousness.

My case study

May 26, 2016 marked my 3rd session of reflexology with a woman named Jane. I had begun acupuncture to address severe abdominal pain that Western medicine had failed to resolve. After 19 years of using a medication (Prevacid), I found myself desperate to find something else that worked. Originally touted as safe, Prevacid and other medications in its class became associated with increased risk of Alzheimer's Disease, osteoporosis, and stomach cancer. While exploring options beyond Prevacid such as acupuncture, I also chose to add reflexology to the regime.

Had I not met and married my Japanese wife, Kaya, ten years earlier, I would not have been exposed to a Japanese healing art, Sei Tai (整体), which Kaya's mother's boyfriend practices. If I had not benefited from Sei Tai I probably would not have been open to trying acupuncture or reflexology. In any event, at the end of my second reflexology session on May 12, Jane inquired into my field of work. When I told her I was a VA psychiatrist, she proceeded to tell me about how her son-in-law, a Gulf War veteran, had committed suicide years earlier and that he had recently "visited her" to say he was finally at peace. Having no patience for allusions to the paranormal, I politely ended the conversation. I vowed that in future interactions with her, I would keep our conversations brief and focused.

My vow of silence did not last. I entered my third reflexology session in an entirely different state. I had been experiencing intense feelings of guilt and remorse for something I had done in the previous week or two. Feeling undeserving and conflicted, I found myself lying on the table unable to relax. I actively summoned self-forgiveness, and in doing so, I came closer to prayer than I ever had in my life. Eventually, I settled on the idea of explicitly instructing myself, "Whatever you experience during this session, it's ok." I now recognize that this statement was an attempt to achieve unconditional acceptance, and I have since learned that self-forgiveness is one of the most common themes in SMEs (Foster, Kahoe, et al., 2018). As it turned out, that was the last fully conscious thought I recall having had before my world was turned upside down.

The next thing I remember was the voice of a woman speaking to me in my right ear. I have since learned, from Dr. Brian Weiss, psychiatrist and author of numerous books including *Many Lives, Many Masters,* that such voices are typically heard through the right ear. The woman's voice I heard was slow, steady, firm, determined, insistent, persistent, hypnotic, and eventually quite annoying. Over and over again, she repeated the same two words: "Tell her. Tell her. Tell her. Tell her ..."

The woman would not leave me alone. Saying "Okay, I'll tell her" was not enough. She seemed convinced I would forget to tell her. It was as if she knew I tend to be forgetful. So she kept hounding me until I was certain not to forget. She wrestled a promise from me to convey the message, and was firm and insistent that I deliver it. I needed her to leave me alone, so I continued to promise sincerely. Finally, the voice ceased.

The body of literature on SMEs shows that spontaneous mediums typically feel "compelled to act or help and want to deliver the right message that discarnates were communicating to them," (Foster, Kahoe, et al., 2018, p. 181). Certainly the woman had compelled me to deliver the message. I also knew the message was for the reflexologist, even though she did not mention Jane by name. Interestingly, the message was not yet encoded in language or even imagery. I only vaguely remember receiving the substantive content of the message. It was like a download of pure information. After the SME, my brain automatically encoded that message into language, but at that stage the only words I recall having heard were: *tell her.*

The next thing I remember happening took place some indefinite time later. I had no sense of time, as I was in a trance. The time distortion I experienced is a common feature of the SME (Corneille & Luke, 2021). I had been in a liminal state, too: not awake, not sleeping, and not dreaming either. In fact, while SMEs can occur in "all states of consciousness," most spontaneous mediums report being in a state that is neither like being awake nor like dreaming (Holden, Foster, et al., 2014). Slightly more awake with my eyes open, I had a vision of a blonde woman all in white, hovering horizontally just beneath the ceiling. She appeared roughly five feet above me and to my right. As if swimming the breaststroke or treading water, and moving in a coordinated but idiosyncratic fashion, her hands were in constant motion. I saw she was attempting to communicate with me, but I could not make out what she was trying to say. As I tried to focus my eyes, the apparition began to move as if to evade my direct gaze. She floated from above my right hip to above my right shoulder, and then above the right side of my head before she finally disappeared through the wall behind me.

Had I been in an ordinary waking state, this might have been a frightening experience. However, my response to watching a ghost float through the wall was simply to close my eyes again. Feeling little to no distress in spite of an anomaly is another common feature of SMEs explored in the case study literature (Foster, Kahoe, et al., 2018).

When Jane finished the treatment, she said, "OK, Michael. Take your time. I'll be outside." She moved to the door. When she touched the doorknob, it was like I received a signal: If I don't act now, I might forget to keep my promise. I suddenly said, "Wait!"

Hearing myself speak jarred me awake. I was so puzzled and incredulous that, instead of uttering, "I have a message for you," I bumbled through the remaining words.

"I have a ... message? ... for you?"

I was bewildered as I heard myself. I knew I hadn't been sleeping. This had not been a dream.

Her response was as unfazed as if I had asked for a glass of water.

Jane: Yes. What is it?

Me: Do you live in a house?

Jane: No, I live in a small apartment. Why?

Me (Squinting my eyes and looking perplexed): Well ... are you making some kind of improvements on your apartment?

Jane: Well, I've been cleaning up, organizing, rearranging, and throwing things away. Why? What's the message?

Remember, the apparition communicated nonverbally. The only words I heard were *tell her.* Yet when I opened my mouth to deliver the message to Jane, the following sentence spilled out: "Whatever improvements you are making on your home, it's ok."

It was very strange delivering a message I had not actually heard—as if the information was encoded until the exact moment I opened my mouth. In his analysis of the mediumship phenomenon, Bernardo Kastrup (2012) confirms that such encounters take place "outside of space-time" and "not in the form of language." Kastrup goes on to say, "It is only after the communication that the perceiver's consciousness retroactively translates the subjective meaning perceived into a storyline and images that make some sense according to ordinary concepts and language." This is exactly what happened to me. The apparition transmitted nonverbal, direct semantic information. To convey the message to Jane required me to encode the information in a language we both understood.

After I said, "Whatever improvements you are making on your home, it's ok," Jane matter-of-factly replied that she had been in the process of throwing away items associated with a deceased lover. She said the work was psychologically difficult, but she was making room for the possibility of a new romance in her life.

At that juncture I responded, "Well, then I have something else to tell you." And I then told her of the apparition.

Jane responded with a smile. "Well apparently we've had a visitor!"

Jane then went on to explain that her former lover was a Native American woman, named Gail Walker, whose indigenous name was Gail Windwalker. She told me Gail had been a psychologist who practiced in upstate New York, where her specialty was grief work. Gail had died eleven years earlier, on June 7, 2005.

I then asked her for a picture of Gail. Jane said she had just the previous evening taken a picture of Gail out of her purse and that she would bring the photo with her on our next visit two weeks later. Now I recognize the coincidence of her having just put the photo away the day before my visit as a synchronicity: one of several that would occur in conjunction with the spontaneous mediumship experience.

The veracity of the apparition was also uncanny. When I returned to reflexology with Jane two weeks later, I said, "You mentioned Gail was Native American. But the woman I saw was blonde." She said, "Yes, that's right. She was 3/4 Irish." This prompted me to request the picture she promised to bring. She said she had forgotten it, and would bring it to the following session. But she never did bring that picture. Sometimes I would forget about it, too. Then, after a couple of months, I asked for the picture again. She told me it was too emotionally painful for her but if I wanted, I could see a picture of Gail by simply Googling her.

I went home intent on looking her up, but I forgot to Google Gail for yet another month. It wasn't until September of 2016 when I finally typed "Gail Windwalker" into a search engine. There she was. The picture I saw was a spot-on match of the woman I had witnessed floating above me. Even her hands were placed over her head in the same position I had seen them. As I read about Gail Windwalker, I learned the reason for her hand positions: she communicated via American Sign Language (ASL), which was what I observed as she was floating above me, trying to communicate with me. I was stupefied.

Spontaneous mediumship and synchronicity

The type of SME I experienced is known as a drop-in. With a drop-in, the apparition appears suddenly and without invitation. Drop-in cases can become a rich source of data on SMEs because they reduce the likelihood of bias; the information that is received during the mediumship would not have been obtained in any other way (Kelly, 2010). A spontaneous drop-in makes the medium a neutral third party; yet the experience is not totally depersonalized. And this is where synchronicity becomes a factor. By triangulating with me, Gail and Jane enabled me to confront my existential beliefs and encouraged me to work through some pressing issues in my personal life.

One of those issues was related to forgiveness and self-forgiveness. Around the time of my SME, I struggled with self-forgiveness. The SME lifted a huge psychic burden from my shoulders while it also allowed Jane to forgive herself for moving on from Gail. In fact, Gail's message to Jane, "whatever improvements you are making on your home, it's ok," echoed what I had told myself at the start: "Whatever you experience during this session, it's ok."

Another personal issue addressed by the SME was my daughter's mental health. Before the Windwalker event, my then 8 year-old daughter had requested to speak with the school counselor to seek help around her profound fear and dread of death— hers and mine. My wife and I were already engaged in the process of searching for a psychotherapist to help her with this anxiety when I had the SME. An abundance of research shows how ADCs mitigate fear of death Corneille & Luke (2021). My SME occurred synchronously with my desire to help my daughter deal with her existential angst.

In addition to highlighting how I might help my daughter overcome her fear of death, the SME highlighted the very same gaps in patient care the VA hoped to address with a short-lived push to integrate spirituality into treatment. My cognitive flip coincided with the VA's initiative to bring patient spirituality into mainstream care. After a year of post-processing and integration, I became more open with patients about spirituality.

Also, I experienced my first SME during bodywork, which is not uncommon (Holden, Foster, et al., 2014). But what was uncommon was for me to seek out complementary medicine for something I had previously relied on only Western medicine to treat. Incidentally, this complementary medicine cured my abdominal pain. If it weren't for my illness, I would not have resorted to alternative medicine; were it not for doing reflexology, I would not have had the vision; if I had not had the vision, Jane would not have received the message.

One day, I opened up to a patient about my newfound interest in consciousness. The patient told me about Stuart Hammeroff, an anesthesiologist who, with Nobel laureate Sir Roger Penrose, founded the orchestrated objective reduction (Orch OR) theory of consciousness. Hammeroff is director of the Center for Consciousness Studies at the University of Arizona. For the past 30 years, the Center for Consciousness Studies has hosted the Science of Consciousness Conference (TSC). I attended the conference in person for the first time in 2022, where a cascade of synchronicities ensued, many of which helped me to integrate and understand my SME. For example, I first met Julia Mossbridge and learned of the AAPS. Research shows SMEs may coincide with synchronistic experiences, such as I have experienced in droves after meeting Gail Windwalker (Corneille & Luke, 2021; Holden, Foster, et al., 2014). This was certainly the case with me.

Interconnectedness and the collective consciousness

Jung's (1985) description of synchronicity as an "acausal connecting principle" shows how anomalous experiences like SMEs can reveal the webbed nature of reality. Anomalous experiences like SMEs can also point to the interconnectedness of all beings, or perhaps even a collective consciousness underlying our individualized identities. For example, I barely knew Jane, and had never even heard of Gail Windwalker before the SME. The SME linked the three of us together—four including my daughter—and also linked me with a slew of people with whom I can now collaborate to understand this and similar phenomena—like the people I met at the TSC conference.

Meeting Gail Windwalker changed my life. The SME prompted me to participate in ongoing research into psi phenomena, taking the experience beyond the personal and into the collective domain. One possible area for future research is in the hypothesis originally postulated by William James that the brain serves as a transducer, rather than a producer, of consciousness. Another area is how we may study synchronicity and psi not as phenomenology but as physics.

References

Corneille, J.S. & Luke, D. (2021). Spontaneous spiritual awakenings: phenomenology, altered states, individual differences, and well-being. *Frontiers in Psychology 12*(2021): https://doi.org/10.3389/fpsyg.2021.720579

Foster, R.D., Kahoe, B.M. & Nardelli, D.H. (2018). Spontaneous mediumship experiences among near-death experiencers: A qualitative study. *Journal of Near-Death Studies 36*(3): 172–184.

Foster, R.D., Lee, D. & Duvall, A.G. (2015). Two cases of spontaneous mediumship of near-death experiencers. *Journal of Near-Death Studies 34*(1): 44–56.

Holden, J.M., Foster, R.D. & Kinsey, L. (2014). Spontaneous mediumship experiences: A neglected aftereffect of near-death experiences. *Journal of Near-Death Studies 33*(2): 67–85.

Jung, C.G. (1985). *Synchronicity: An acausal connecting principle.* London: Routledge.

Kastrup, B. (2012). *Apparitions, ghosts, and mediumistic communications.* Retrieved from: https://www.bernardokastrup.com/2012/09/apparitions-ghosts-and-mediumistic.html

Kelly, E.W. (2010). Some directions for mediumship research. *Journal of Scientific Exploration 24*(2): 247–282.

Peat, F.D. (n.d.). *Time, synchronicity, and evolution.* https://paricenter.com/library-new/c-g-jung/time-synchronicity-and-evolution/

Seda, D.A. (2023). Mediumship and mental health. *Journal of Conscious Evolution 20*(2): https://digitalcommons.ciis.edu/cgi/viewcontent.cgi?article=1142&context=cejournal

Chapter Twenty

Meeting My Mentors

KARALEE KOTHE, PHD CANDIDATE

Karalee Kothe has been a student of consciousness since she was a kid. The beauty of nature, the joy of music and dancing, the exploration of knowledge, and the enduring love of her family have been her guiding lights on her path. She obtained a master's degree from the Spirituality Mind Body Institute at Teachers College, Columbia University, and she is currently a clinical health psychology PhD student at the University of Colorado, Denver. Her research interests presently include understanding terminal lucidity, the sudden return of energy and clarity before death, and examining the intersection of meaning, spirituality, and health.

Synchronicity #1

When I was 21, I had never heard of Jung, which now strikes me as terribly ironic because I was a psychology major. It seemed the psychology curriculum at my college wanted to prevent us students from getting too inspired or too disturbed. And yet, all I longed for in my education experience was to be both inspired and disturbed. At this point in time, I was nearing the end of my undergraduate degree in psychology and aside from a "Death Becomes Us" class and work in the health psychology lab, I was feeling deeply unsatisfied with the lack of depth in my learning experience so far. I hadn't felt transformed or enlightened as I had expected when I began college three years prior.

Since my curiosity wasn't being satiated within school, I began to look outside the curriculum. I came across some books on quantum physics, and one book in particular, the *Self-Aware Universe,* by Amit Goswami (Goswami, 1995). I immediately felt something stir within as I began to understand how spooky and magical the implications of quantum physics are—as well as what this could mean for consciousness—that it is primary rather than an epiphenomenon. Nothing in my psychology courses stirred me like this book did.

While I had never been exposed to these ideas before in my education, something about them felt so familiar, like a homecoming, like everything was sliding into place, and I felt an aliveness that lit up my mind. Perhaps this is why I often felt so disillusioned by education as it was missing this tangible sense of aliveness. And so, when my boyfriend at the time asked me to visit him over winter break, I of course brought Goswami with me. My boyfriend lived in Portland then, and I had never been there before. I read Goswami on the plane completely engrossed in every page. By the time I arrived in Portland, I was about half-way through the book.

The next day I went with my boyfriend to the famous Powell's City of Books. Immediately, I was over-stimulated by the sea of books. I began to wander aimlessly, all the while, still thinking about Goswami, which of course I had brought with me. It was hard to be interested in another book when I was so in love with this one. After skimming a dozen different titles, I decided to give up and go and read Goswami in the café.

I started walking towards the café until one book really caught my eye—much more than the hundred other books I had just walked past. The book was *Synchronicity*, by Carl Jung (Jung, 1952). Not sure why this book caught my attention as the cover wasn't particularly attractive and I was not familiar with the word, "synchronicity." Yet, inexplicably, I paused to pick it up and I looked through a few pages. I quickly lost interest and put it back on the shelf. Back to Goswami ...

I sat down just twenty paces away from the synchronicity book and began to read. I turned the page, and the very next paragraph began with a reference to "Carl Jung's Synchronicity..." I gasped out loud and a sense of numinosity flooded my being. Now, Jung had my attention. I had an explanation of synchronicity in the next few lines of text that described exactly what had just happened to me in live-time.

In that moment, the trickster nature of the universe was unmistakable. And, I thought, Jung must be one of the great tricksters. My mind raced through the improbabilities of the perfect timing. Had I read one page more on the plane I would have been less impressed with the order of events. Or, had I not flown to Portland, had I not started dating again, had I not found *Self-Aware Universe,* which all are very likely alternative timelines, I would not have experienced this synchronicity, and I would not have met Carl Jung. Little did I know then that this was the beginning of what I imagine to be a life-long mentorship.

Three years later I found myself in Manhattan completely engrossed in a lecture by a Jungian analyst, and feeling that sense of aliveness again. Then, I read *The Red Book* and it was truly a spiritual experience for me as I felt Jung's presence so strongly through those pages—and even though I was living alone in strange city that felt truly alien to me, Jung brought the world back alive (Jung et al., 2009). I could feel the magic, even in NYC. I realized I had come full circle when our professor asked us one day, "Who has an example of synchronicity?" and I smiled to myself, realizing my synchronicity about synchronicity is one of my greatest treasures.

Synchronicity #2

I first read Viktor Frankl's, *Man's Search for Meaning,* when I was maybe 14 or 15 (1985). Frankl's experience was a welcome friendship for me at the time as I often found myself in and out of existential crises and general depression with being over-achievement oriented when I really should have been still experiencing my childhood. My nana had given me the book as we tended to enjoy similar topics and we discussed the book together.

Years later in 2021, I found myself once again in a bit of an existential crisis—I had just graduated college and the COVID-19 pandemic had just begun. Basically, two rugs had been pulled out from under my feet at the same time—first, the cozy container that is the undergraduate experience and second, the norms of society and the complete shock of a global pandemic. I certainly was not prepared for this second wave of groundlessness at this stage in my life. What is a freshly graduated person to do in such a world? All I knew at the time was that I needed more—more education, more expanding, more learning.

And after my introduction to Goswami, I realized it was consciousness that I wanted to dedicate my life to, whatever the heck that meant. I was startled that no one was talking about consciousness, it seemed to hold both the questions and answers to all our problems, and I couldn't stop thinking about it. Thank God I was born into the information age because I sought more everywhere—I read articles, emailed professors and authors whose work I found fascinating, and started attending online conference after online conference related to post-materialist consciousness topics.

Even though I wasn't in school, I made my life into school again. In this series of online conferences and summits, an old friend began appearing again, Viktor Frankl. I was amazed how often he was part of a presentation or discussion at seemingly unrelated conferences. So, I figured I ought to read *Man's Search for Meaning* again. At this point, I was reading Frankl, listening to people talk about Frankl, and discussing Frankl with my nana again, when I was watching yet another presenter give a Frankl overview. At the top of one of his slides he had listed Frankl's lifetime in years, like this: "1905–1997." The year, 1997, got my attention, because that was the year I was born.

I felt this overwhelming urge to google the exact day Frankl died, and low and behold, he died September 2nd, 1997, the exact, same day I was born. This incredible coincidence felt incredibly personal to me, at a very meaningful point in my life. Was Frankl trying to pass the baton to me, to carry on his work with meaning? Although I had known Frankl for ten plus years, and I had never had the inkling to look up the day he died. Instead, I discovered we share such a meaningful date at a time when I was feeling completely groundless during my gap year, and during the global pandemic. Life meaning had always been salient to me and was beginning to inform my approach to psychology and to my graduate degrees.

And so, discovering the creator of logotherapy died the day I was born was pivotal for me and my life path. Not surprising then, a year later, I was interviewing for a PhD program, a program I chose because of the lab called, "Cardiovascular Health and Life Meaning." (I intuitively knew "meaning" was a, yes!) I had to go somewhere that understood the importance of meaning! I interviewed with the director of the lab and I had to ask him, "Are you familiar with Viktor Frankl?" He lit up, and we proceeded to spend the rest of the interview discussing Frankl. I got into the program. Then in the second year, I took his existential psychology course and we spent a few weeks on Frankl again. My dear Frankl, I think this is the beginning of a beautiful friendship.

What does it all mean?

Jung and Frankl both appeared in my life through a series of synchronicities precisely when I needed them the most. At times when I felt lost, unguided, or in the midst of a developmental depression, they emerged to be my teachers, my guides, and to give me hope that there is something more going on here than what I had been taught in school. Both mentors taught me their most important discoveries of their careers/lifetimes through lived experience of their theories. And both of their appearances set off a series of chain reactions that dramatically shaped my career and life decisions.

I would never have gone to a master's program in NYC if it was not for Jung, and I would never have gotten into a PhD program right in my own backyard of Colorado if it was not for Frankl. More importantly, I may have fallen into a deeper disillusionment and disenchantment with the world without their intervention. Emerging adulthood is really a break down or break open kind of time; especially alongside the COVID-19 pandemic, I felt I was at higher risk for disillusionment. While I did take action towards a greater purpose that was more in alignment with my life in both cases, I did not have any idea where my actions would lead me or how these events would unfold. My actions were answered with outer guidance exactly when I needed the "God nod," to confirm the tentative steps I was taking off the beaten path.

Beyond my own path, these synchronicities pointed me towards the truth of the nature of reality. The Frankl synchronicity implied the inherent meaning in death. Understanding what death tells us about life has been a growing interest of mine ever since I was 18ish. The Frankl synchronicity revealed to me that dying, in and of itself, is an incredibly meaningful, numinous, spiritual event because the death of one person may have ripple effects on many, many people.

For instance, if in my own small example, i.e. Frankl's death, a person I never knew, a person whose life I hardly overlapped with, changed the course of my life, who is to say the way in which millions and millions of deaths effect millions and millions of lives? We may only have a glimmer of how much our life is governed not only by our own death, but also by the deaths of others. But what we do know is that our shared mortality is perhaps our greatest common ground because we all have loved someone who has died, and we all must face our own death.

Although American culture makes every effort to ignore death and its inevitable presence in our lives, the trickster nature of the universe may be pushing us all towards the realization that it is death that makes life possible. Without death, life may be meaningless, and a life without meaning is no life at all. Frankl's life's work certainly showed that lack of meaning is the deadliest disease to humankind. Of course, Frankl's life has touched perhaps millions of lives, but who would have guessed the way it would impact mine?

And why would we share the date for the two most pivotal events of a human's existence? To me, this implies that perhaps, death and birth are one in the same, two sides of the same coin. The full circle with Frankl, is that with death, comes birth. And because life and death are so inexplicably in marriage with one another, meaning must be inherent, must be innate in the fabric of the universe. But this meaning only arises because death is the ultimate horizon of human existence (see Heidegger's *Being and Time*, 2010). It becomes hard to separate what comes first; is it birth, death, or meaning?, rather, they seem to come all at once, in a mysteriously entangled way; perhaps quantum physics can answer that mystery—you can't get one without getting the other two.

Of course, Goswami's book holds some key answers to the truths I am pondering, and so he must inform my conclusions as well. He writes, "... our observation collapses the quantum wave packet to localized particle," (Goswami, 1995, pp. 45). In other words, conscious intention gives direct form and meaning to an otherwise cloud of unmaterialized probabilities.

If I described the above events without a primary actor or author of them, i.e. myself, they would appear quite meaningless. They would be just a book, a bookstore, a date in time, and another book. What links the random events together is my conscious intention to realize the meaning inherent within the events. Furthermore, my inner world is necessary to link together the seemingly random outer world.

And if at the quantum level, all of reality awaits in probabilities of quantum wave packets, then some form of agency or action is required to focus probabilities into actual realities. I perceive this as a kind of tennis match with the universe. Inner events play one side, (which includes the individual person) and outer events play the other side, (which includes everything that is perceived to be outer, or not-me).

But who or what the other player is remains a mystery. Both players must participate for the game to exist. If I stop playing tennis, the game no longer exists and tennis balls keep bouncing over a net to no answer, no focusing intention to respond to each ball, to make the ball more than a random bouncing ball, to make it a match. The match transcends both players, it becomes the third thing, the "Transcendent Function," as Jung called it (Jung, 1957). Therefore, in the tennis match, the meaning of the match, which is an idea, arises and supersedes the participation of each player.

Thus, a meaningful universe requires participation from conscious agents to bring into being something more than the inner and outer worlds combined: a synchronicity. Likewise, my engagement with life naturally led meaning to arise—meaning being the focused intention of a mere possibility that was yet to exist—just as conscious observation of quantum objects leads to a collapse of the wave function. As such, birth and death could also be examples of probabilities becoming realities, and so the reverse may also be true, that humans are unmaterialized wave packets of pure potential until the conscious intention of the universe looks upon us and collapses us into this human reality. It takes two to play the game.

In my life examples, when my belief in the inherent meaning of the universe was threatened, the universe answered in the most meaningful way it could: with synchronicities. Synchronicity is a language that a lost human can understand because it is not thinking about meaning or learning about it, rather, synchronicity is the lived experience of meaning. It is the difference between thinking about love and actually loving someone directly. The universe must care when we stray into meaninglessness and has devised this clever trick of synchronicity for restoring our full-hearted faith in meaning.

Thus, there must be a loving, guiding intelligence interwoven in the fabric of reality, with built-in guideposts to bring us back to truth when we are lost. Because when we forget the truth of reality, we forget who we are. And I think it is only in this state of amnesia when a person could believe there is no outer or inner meaning, and all events, lives, and deaths are isolated, random events. When in reality, simply existing as a human is automatically meaningful as an individual's existence is a focusing of what would be only probabilities.

And because a human life is finite, the lack of infinite outcomes is the precursor required for meaning to arise. If you change one variable in my synchronicities described above, the meaning dissipates. The specific variables in my specific and limited circumstances were the focus required for meaning to materialize. However, this meaning would never have emerged in the first place if I hadn't been chasing after what made me come alive.

This "aliveness" is truly one of the sacred ingredients for making synchronicity soup. Meaning arose from the events above because I was following my bliss. I pursued with passion whatever it was that would stir that thing with feathers that lives within the heart space. And now, I pass the mic to Joseph Campbell to take us home:

> *"Follow your bliss. If you do follow your bliss, you put yourself on a kind of track that has been there all the while waiting for you, and the life you ought to be living is the one you are living. When you can see that, you begin to meet people who are in the field of your bliss, and they open the doors to you. I say,*

follow your bliss and don't be afraid, and doors will open where you didn't know they were going to be. If you follow your bliss, doors will open for you that wouldn't have opened for anyone else."

References

Frankl, V. E. (1985). *Man's search for meaning.* Simon and Schuster.

Goswami, A. (1995). *The self-aware universe: How consciousness creates the material world.* Penguin.

Heidegger, M. (2010). *Being and time.* Suny Press.

Jung, C. G. (1952). Synchronicity: An Acausal Connecting Principle. Collected Works (Vol. 8). Princeton, NJ: Princeton University Press.

Jung, Carl G. "The transcendent function." (1957).

Jung, C. G., Shamdasani, S. E., Kyburz, M. T., & Peck, J. T. (2009). *The red book: Liber novus.* WW Norton & Co.

Chapter Twenty-One

Synchronicity as the Yoga of Absolute Knowledge

RICKY DERISZ

Ricky Derisz is a writer, speaker, and host of the *Mind That Ego Podcast.* Inspired by his battle with mental illness and subsequent awakening, Ricky's independent research and practice of self-inquiry synthesizes an expansive network, including depth psychology, spirituality, philosophy, and science, to support a cohesive model of reality that allows for a life-enriching relationship with the cosmos, one of enchantment, mystery, and mysticism.

My relationship to synchronicity has evolved across three stages. Initially, on rare occurrences of coincidence, I'd be dismissive and mildly freaked out. I was "scientifically-minded," atheistic, and intellectual. If coincidences were shared (suggesting a song a friend was about to play was eerily common on car journeys), we'd pause, hold bemused eye contact, we'd say, "that's weird," forget about it, and move on with our lives.

In my early 20s, after years of chronic depression and anxiety, an ever-darkening dark night of the soul manifested as paranoia and psychosis, catalyzed by synchronicity's shadow, "tendencies towards narcissistic distortion, by which random or peripheral events are continually transformed into signs of an egocentric universe" (Tarnas, 2006, p. 9). Those random and peripheral events included auditory hallucinations of persecution.

Eventually, after years of meditation and self-inquiry, a critical mass occurred. Synchronicities became the hallmark of an exhilarating awakening. To paraphrase the *Bhagavad Gita:* "That which is like poison at first, but nectar in the end." Whereas previously my mind was dominated by neurotic thoughts, a shift occurred. Ideas, insights, and revelations filled my consciousness. I was able to intuitively understand complex theories from physics, philosophy, and psychology.

Despite this exalting breakthrough, I judged knowledge as a hindrance. Spirituality is not conceptual, they say, and I assumed connecting to God was through silence, the "language God speaks," as Rumi claimed. But in Eastern philosophy, there are multiple pathways to yoga (union): Karma Yoga (selfless service); Bhakti Yoga (devotion); Raja Yoga (concentration); and Jnana Yoga (knowledge).

The etymology of meaning, *meninge,* is the "act of remembering." It's fitting, as the spiritual path is often described as remembering who you truly are. Here I will focus on meaningful coincidences related to Jnana Yoga, and explore how synchronicity may light the path of union with God through the intellectual domain, by "matching" intuitive insights with pre-existing bodies of work.

The prism and the scarab

My first intense experience of synchronistic knowledge occurred during a meditation session. A vivid image of white light refracting through a prism appeared in my mind's eye. I had an intuitive understanding of what this represented. Instinctively, I searched online for the prism as a spiritual symbol, unconcerned about the "insanity" of searching for an abstract image that surfaced inwardly.

I found an excerpt from *You Can Find Inner Peace* (George, 2013). I'd not heard of the book or author before. As I read the passage, I knew the message was for me. Words formed like apparitions from my intuitive understanding. This was distinct from the constriction of egocentric distortions; it was accompanied by expanded energy, a heartfelt affirmation, a transmission between the words and me:

> If we retrain ourselves to open up to the source of spirit from whom we came and to whom we will return, we will feel the rightness of this relationship, and we will also feel its joy. The profundity of spirit is symbolized by white light — a simple pure beam that draws together every color under the sun. White light is as invisible to us as the air we breathe. Yet we can see it when it is separated into its constituent colors in a prism or rainbow. Similarly, we are directly aware of the source when it is refracted through the prism of our intuitive experience. (p.41).

If you've ever read a synchronistic message, you'll know it lands differently. I enter a trance-like state, where I'm primed for a cosmic whisper. Time slows as the message arrives, full to the brim with emotional resonance and transcendent beauty. No matter the source, be it graffiti seen on a Monday morning walk in Berlin, a billboard or book, a slogan on the clothing of a passer-by, it's as if my soul emphatically projects messages onto the canvas of my experience.

On one occasion, while planning an event on synchronicity, I first read Jung's story about his client's dream and the scarab beetle, and considered what example to use in the presentation. I took a break, logged into Facebook, and the first post in my feed was an image with the headline: *What are our ancestors trying to tell us?* It was accompanied by a quadrant, with four Egyptian symbols compared to nature—including a scarab beetle next to a human skull.

My mind countered the coincidence, thinking up ways my online searches must've been tracked by the algorithm. Yet Jung's story appeared in an article with unrelated terms. I checked the time of the post — it was published minutes before, the exact time I was reading Jung's reflections on miracles that crack the veneer of rationality. I obliged, and added the screenshot to the presentation, with the caption: *here is your scarab beetle.*

The headless Buddha

What happens when the initial symbol crosses the threshold into the physical world? Last year, prior to a trip to the UK, I woke up conscious of my decision to take Alan Watts' *Wisdom of Insecurity* with me. As I reached for the book on my bookshelf, I knocked one of three Buddha ornaments onto the floor. Its head came clean off. After a pregnant pause where me and my partner looked at this as a literal bad omen, we both laughed. It was one of those moments.

My intuitive sense was that this wasn't an accident, or chance, but the headless Buddha represented insight. Later that day, as I was traveling to the airport in a taxi, a song came on the radio. The lyrics nudged me through the speaker. "Though I can't hide from the truth, I lose my head over you," by Freya Sky. Just as I was getting worried this was a bad omen, I open *The Wisdom of Insecurity* during the flight, and read Deepak Chopra's preface:

> Watts has already introduced the most basic Buddhist stance: sober examination of what lies before you, leaving aside all assumptions. By holding onto this sense of openness, we can find all truth in ourselves. The promise, as held out here, echoes what saints and sages have taught in every

> wisdom tradition. Where Buddha refused to answer questions about the existence of God, Watts is more inclined to smash idols. (2021, p. 4)

A few weeks before, I wrote an article on spiritual narcissism, which referenced a Zen koan, reportedly told by Linji Yixuan: "If you meet the Buddha on the road, kill him." As with the prism, these words fleshed out my initial intuition and exploration of the theme of iconoclasm, as if my article was coming alive, animated through the lived symbol of an idol smashing on the floor.

Although synchronicities "just happen," once they've occurred, there seems to be an interactive quality, a treasure hunt, as if interpreting a dream, where additional meaning is revealed through active imagination. I wanted to find out more about the symbol of the headless Buddha. What else was waiting to reveal itself? Nothing carried an emotional charge or signified deeper meaning, until I discovered a talk by Watts. The title stood out like seeing your destination pop up as the next stop on public transport. It's called *Swimming Headless.* I started listening:

> Most people move in constant opposition to themselves because they are afraid that, if they don't, something awful will happen. See, it's so easy to bend your arm. But if you somehow felt that, in bending your arm, you might make a mistake – you know, instead of bending it you might hit yourself – then these two muscles get nervous and they fight each other to be sure that everything happens alright. That's anxiety, you see?

The surreal specificity of bending an arm and making a mistake jolted me into paying extra attention. Again, time slowed, consciousness shifted, as I was witnessing the trajectory of an arrow of thematic meaning traversing multiple planes, ornaments, books, songs on the radio, audio-recordings of lectures from years-gone-by. Watts continues:

> In this sense, the invisibility of one's head – almost the not having of any head at all – is the secret of being alive. To be headless, you might say. To have no head, in just the sense I'm talking about, is our way of talking about the Chinese expression mushin, 'no-mind'.

Interpretations of knowledge and free will

In the same talk, Watts explains how the conventional approach to knowledge is like an encounter where we "face facts," as if "somehow the knower and the known came from two completely different worlds." But he suggests another perspective. "Rather than a collision between two wandering bodies in space, knowledge is much more like the expansion of a flower from the stem in the bud, where the opposite points of the flower are the knower and known."

The model of the conscious and unconscious mind helps to understand how this flowering relates to synchronicity. Von Franz (2013, p. 211) describes the archetype of the Self as existing "beyond the consciously realized flow of life." Its organizing principle transcends mind and matter, sometimes referred to as dual-aspect monism, where thoughts and feelings and atoms and quanta orbit a substrate of meaning.

Mentioning a song a friend was about to play could be explained by parapsychology. But the example of the Headless Buddha, if orchestrated by meaning, consists of an unfathomable degree of fine-tuning. The placement of the book, the precise movement to knock the ornament on the floor at an angle to make a clean break, the timing of reading *Wisdom of Insecurity,* its thematic relevance of smashing idols, Watts' using the metaphor of bending arms and making mistakes...

In *Synchronicity* (1985) Jung cites Camille Flammarion's similar coincidence between knowledge and the physical world. While writing a chapter on wind-force for *L'atmosphere* (1888), a gust of wind blew his papers out of the window "carrying them off in a sort of whirlwind among the trees." When it started to rain he decided not to retrieve them, and yet, miraculously, the printed version contained the missing pages. Flammarion's porter had walked by moments later, presumed he'd dropped the pages, dried them, and delivered them to be printed.

Aptly, the printed version of *L'atmosphere* included what has since become known as the Flammarion engraving, which depicts a pilgrim peering beyond the veil of duality, into the unified world, or *unus mundus.* It's commonly used to symbolize synchronicity, echoing Medieval alchemist Gerhard Dorn's description of the spiracle, from the Latin *spiraculum aeternitatis,* the breathing hole or window into eternity (Brown, 2021).

Such experiences imply that "all the universe conspires," (Coelho, 1988) where even a gust of wind is an animating force playing its role. Could my unconscious have orchestrated a fine-tuned chain of events, without my conscious awareness, to create the symbol of the headless Buddha? Could the idea to read *Wisdom of Insecurity* have been retrocausal, a message back in time after the ornament was knocked to the floor?

And if my actions and instinct are part of an orchestrated manifestation of meaning, do I have free will? Was philosopher Friedrich Wilhelm Joseph von Schelling accurate when he wrote:

> "The I think, I am, is, since Descartes, the basic mistake of all knowledge; thinking is not my thinking, and being is not my being, for everything is only of God or the totality." (1856, p.148)

Divine intervention and intuitive intellect

I don't see humans as automatons, but give credence to Schelling's notion that thoughts interface with God, explaining why knowledge can lead to mystical union. Not through memorizing lifeless concepts, but through an embodied connection to sacred knowledge, in an act of remembering. That implies that knowledge is a thing in itself, what thirteenth-century theologian St. Thomas Aquinas describes as *positively immaterial* objects that exist independent of worldly matter.

According to Aquinas's Theory of Knowledge, unaided human reason is unable to comprehend the positively immaterial, which is beyond the intellect's usual parameters. However, Aquinas suggests that positively immaterial objects have agency, similar to archetypes, and that direct knowledge was possible with guidance or revelation from God (Kerr, G.,n.d.). In other words, a form of divine intervention supports an individual to perceive knowledge in ways usually inaccessible.

The etymology of idea, the Latin *idein*, means "to see," and is defined as a "concept of a thing in the mind of God" (the universal) or a "mental image or picture" (the personal). This mirrors Aquinas' theory. But what facilitates revelation? Western philosophers, such as Kant and Hegel, define *intuitive* intellect as transcending the mind's usual limits (Westphal, 2009). Similarly, Spinoza's intellectual love of God, *amor Dei intellectualis,* occurs when the "mind passes to the highest state of perfection that is available to it." (Dutton, n.d)

Intuition as higher mind appears in Eastern philosophy, too, accessible when thought is stilled through meditation. Patanjali's Yoga Sutras note that "by intuitive insight everything becomes known." (Vasquez, 2021). In Hindu sanskrit, *pratibhā*, is "intuitive and creative genius," similar to the ancient Greek belief that creative inspiration was a bridge to the divine (Hiemstra, 2017).

As Aquinas claimed unaided reason is unable to grasp the positively immaterial, Jung notes how the "unaided intellect" struggles to comprehend wholeness. In referencing the "intuitive technique for grasping the total situation" demonstrated in the *I Ching and Book of Changes,* he writes: "Unlike the Greek-trained Western mind, the Chinese mind does not aim at grasping details for their own sake, but at a view sees the detail as part of a whole." (1985, p. 34).

For Jung, adequate judgment requires support from sensing and intuition, as *a priori* knowledge, or absolute knowledge, is comprehended via inner perception and not cognition, as if revealed from a transcendent source. Unsevered from the whole, the intuitive intellect can comprehend absolute knowledge prior to its formation into words, ideas, or concepts. Returning to synchronicity, I sense meaningful coincidences "match" intuitive knowing with already-formed concepts or ideas, through a common source of absolute knowledge.

Like a scholastic match-maker, the dual-aspect monad attracts works from others who created from a similar intuitive knowing. Synchronicity breathes life into the process through affirmation and emotional resonance, as if remembering, not learning anew. Phrased another way, synchronicity endorses truth by highlighting the footsteps of those who opened the spiracle to the timeless domain of absolute knowledge and attempted to explain what they found.

Human translations of these experiences aren't perfect forms. But when attempts are made to translate, the concepts and words are rooted in absolute knowledge, with inherent meaning encoded into the work, in a way that transcends mere information or rhetoric. This encoded meaning facilitates synchronicity, as if certain works have morphic resonance (Sheldrake, 1981), and once an intuitive knowing is activated (or remembered) in an individual psyche, the cosmos conspires to draw relevant works towards the person, should their thirst for knowledge be relevant and meaningful.

Perhaps in the mind of God, absolute knowledge wants to surface, it wants people to attempt to translate as best they can, as part of an ever-evolving cosmic consciousness. In Hindu thought, *pratibhā* is also poetic intuition, the ability to interpret and create symbolically. Do creative geniuses—musicians, philosophers, artists, poets, scientists—translate absolute knowledge so lucidly their creations can awaken others, pulling them into the timeless space, the *unus mundus?*

It's human nature to experience a shift in perception having listened to music, observed great art, read literature or philosophy, or watched films that seem to portray forgotten memories, hidden truths. Is this why Einstein (1935) claimed pure mathematics the "poetry of logical ideas", or why quantum physicist Niels Bohr said that when it comes to atoms, "language can be used only as in poetry" (Heisenberg, 1981). Is absolute knowledge a cosmic poem, recited by God's small still voice?

Implications for sanity

In the Katha Upanishad, the Jnanic path is compared to a razor's edge (Whitney, 1890). Because the intellect seeks realization by transcending itself, it runs the risk of confusion. Distortions of synchronicity are a trap in waiting. That includes egocentric interpretations of coincidence or absolute knowledge, delusions of grandeur from identifying with archetypal realms, divine madness or spiritual emergency.

Materialist culture is quick to label all synchronicities as apophenia, which psychologist Claus Konrad used to describe perceiving meaningful connections between unrelated things (Mishara, 2010) similar to what executive director of the Skeptics Society, Michael Shermer calls patternicity, "the tendency to find meaningful patterns in meaningless noise" (2008). Both are valid concepts, but defaulting to them as a catch-all dismissal is absurd.

A better bet would be to discern the razor's edge. Carefully defining "unrelated" is essential, as is exposing the assumption that only causal relationships are valid. How many insights would've been labeled meaningless noise by those tone deaf to the rhyme and rhythm of the cosmos, echoing the metaphor often associated with Friedrich Nietzsche: "Those who were seen dancing were thought to be insane by those who could not hear the music"?

It took dedicated inner work to cleanse my doors of perception, to discern between projection and true synchronicity, madness and divine revelation. The process is ongoing, and even now, I have to be careful to not fly too close to the sun or draw cognitive conclusions from synchronicities. But the overall repercussions have been profound, and had I dismissed coincidence as apophenia or patternicity, I would never have turned that poison to nectar.

Memory of previous synchronicities give me resilience through challenging times, and demonstrated that I wasn't mad, but misdirected. Synchronistic knowledge has made my vocation as a writer a communion with the divine. When I feel disconnected or isolated, creativity is a pathway to the transcendent, coincidences in words and concepts, ways for me to remember.

In Buddhism, there are said to be 84,000 Dharma Gates, a metaphor pointing to the many pathways to enlightenment. Jung writes in *The Red Book:* "If one accepts the symbol, it is as if a door opens leading into a new room whose existence one previously did not know" (2012, p. 310). I wonder if the prism appearing in my mind's eye during meditation was an invitation to enter a new room, a room of absolute knowledge and enchanted information?

Beneath every disguise, every form synchronicity takes, from gentle reassurance in moments of suffering, to inspired revelation, to cosmic winks of wisdom expressed through ornaments falling off of shelves, is the underlying truth of unity, aspects of the profound white light of spirit we become aware of when refracted through the prism of our intuitive experience.

References

Coelho, P. (1988). *The Alchemist* (25th Anniversary Edition). Harpercollins Publishers.

Dutton (n.d.) *Spinoza, Benedict De | Internet Encyclopedia of Philosophy.* (n.d.). https://iep.utm.edu/spinoza/

Einstein, A. (1935, May 4). Emmy Noether: NY Times Obituary [Review of *Emmy Noether: NY Times Obituary]. New York Times.*

Friedrich Wilhelm Joseph Schelling's Sämmtliche Werke, ed. K.F.A. Schelling, I Abtheilung Vols. 1–10, II Abtheilung Vols. 1–4, Stuttgart: Cotta, 1856–61

George, M. (2013). *You Can Find Inner Peace.* Watkins Media Limited.

Heisenberg, W. (1981). Physics and beyond. Allen & Unwin, June.

Hiemstra, G. (2017). Pratibha, Pratibhā, Prātibha: 15 definitions. www.wisdomlib.org. https://www.wisdomlib.org/definition/pratibha

Jung, C. G., Henderson, J. L., Marie-Louise Von Franz, Aniela Jaffé, & Jacobi, J. (2013). *Man and His Symbols.* Bowdon, Cheshire, England] Stellar Classics.

Jung, C. G., Sonu Shamdasani, Ulrich Hoerni, Kyburz, M., & Peck, J. (2012). *The red book = Liber novus : a reader's edition.* W.W. Norton & Co.

Jung, C. G. (1985). *Synchronicity—An Acausal Connecting Principle.* Taylor & Francis Ltd.

Kerr, G. (n.d.). *Aquinas: Metaphysics | Internet Encyclopedia of Philosophy.* Internet Encyclopedia of Philosophy. https://iep.utm.edu/thomas-aquinas-metaphysics/

Laurence Browne (2021) *The Flammarion Engraving and its Symbolic Potential, Psychological Perspectives,* 64:4, 562-580, DOI: 10.1080/00332925.2021.2044173

Mishara, Aaron (2010). "Klaus Conrad (1905–1961): Delusional Mood, Psychosis and Beginning Schizophrenia". *Schizophr Bull.* **36** (1): 9–13. doi:10.1093/schbul/sbp144

Mukundananda, S. (n.d.). *Chapter 18, Verse 37—Bhagavad Gita, The Song of God—Swami Mukundananda. www.holy-Bhagavad-Gita.org.* https://www.holy-bhagavad-gita.org/chapter/18/verse/37

Sedgwick (Ed.), *The Reception of Kant's Critical Philosophy: Fichte, Schelling, and Hegel* (pp. 283–305). chapter, Cambridge: Cambridge University Press.

Sheldrake, R. (1981). *A New science of Life : The Hypothesis of Morphic Resonance.* Rochester, Vt. Park Street Press.

Shermer, Michael (2008). "Patternicity: Finding Meaningful Patterns in Meaningless Noise". *Scientific American.* Scientificamerican.com. **299** (6): 48. doi:10.1038/scientificamerican1208-48

Tarnas, R. (2006). *Cosmos and Psyche : Intimations of a New World View.* Plume.

Vasquez (2021). *3.34 Intuitive insight.* https://simple-yoga.org/3-34-intuitive-insight/

Watts, A. (n.d.). *Swimming Headless* [Transcript of Swimming Headless]. https://www.organism.earth/library/document/swimming-headless

Watts, A. W. (2021). *Wisdom Of Insecurity: A Message for an Age of Anxiety.* Rider.

Westphal, K. R. (2000). *Kant, Hegel, and the Fate of "the" Intuitive Intellect. In S. Sedgwick (Ed.), The Reception of Kant's Critical Philosophy: Fichte, Schelling, and Hegel (pp. 283–305).* Cambridge: Cambridge University Press.

Whitney, W. D. (1890). *Translation of the Katha-Upanishad. Transactions of the American Philological Association* (1869–1896), 21, 88–112. https://doi.org/10.2307/2935810

Chapter Twenty-Two

How a Neuroscientist Came to Embrace the Reality of Acausal Synchronicities

LALEH K. QUINN, PHD

Dr. Laleh Quinn has been a research faculty member in the Cognitive Science department at the University of California San Diego for 25 years. She obtained her Ph.D. in Philosophy of Mind, Cognitive Science, and Cognitive Neuroscience from the University of Arizona. Her research has ranged from neurophysiological recordings of brain activity in awake behaving animals to understanding the roots of empathy and compassion. She has taught mindfulness and meditation to students and faculty at UCSD and is currently devoting much of her time to writing about her experiences of expanded reality beyond the confines of materialism.

In Jung's seminal work, "Synchronicity," he argues for a fourth organizing principle—alongside space, time, and causality—at work in the universe. As Jung experienced himself, there are some events experienced by humanity that cannot be explained on the basis of the first three principles, and thus there must be a fourth. This fourth is what Jung referred to as "synchronicity." He defined it as an "acausal" principle, in that it did not seem to be governed by causation as normally defined within the standard scientific model.

Jung describes several instances of synchronistic experiences that could not be explained through physical causation. One of the most famous of these examples is that of the Scarab: while tending to what he refers to as an overly rational patient in his office, who had yet to break through to the deeper reality of her psyche, she related a dream she had the previous evening about a piece of jewelry in the shape of a golden scarab. Just at that moment, a scarab with green and golden hues flew into the office. Jung famously stated: "here is your scarab." The patient was immediately lifted out of her narrow vision of the world.

This example is noteworthy, as it seems to violate what may be acceptable as normal coincidence. Coincidences happen all the time, but they vary in their probability of occurrence. What lies at the heart of Jung's insistence that at least some coincidences cannot be explained by mere chance is a seeming violation of such probabilities. It may be a coincidence that my friend and I both decide to wear red shirts and dark jeans on the same day; however, discussing a dream of a golden scarab and immediately witnessing the appearance of a golden scarab—especially if those appearances are extremely rare, as they were in Jung's environment—seems to point to another level of 'coincidence.' For Jung, these are the true synchronicities. There is something in this conjoining of events that goes far beyond our normal understanding of the way the universe works; far beyond normal chance occurrences governed by the laws of physical causation. After all, the discussion of a dream certainly did not *cause* the scarab to arrive at the window. Moreover, this synchronicity was a highly *meaningful* one for the patient.

Many of us have had these types of extraordinary experiences in our lives, but we tend to either pass them off as normal coincidences or, at best, delegate them as anomalies that must have a proper scientific explanation that we don't yet comprehend. Experiences such as thinking of someone not thought of in years and having them call the next day have probably happened to most of us, but if you're like me—at least the former me—I would have a twinge of excitement about coincidences like that and then promptly forget about them. For Jung, however, attending to such synchronicities goes far beyond feelings of excitement. To him, doing so provides us access to one of the most important forces of the universe, and to an understanding that, at the heart of everything, is *meaning*. As he states in *"The interpretation of Nature and the Psyche,"*

> Synchronicity ... means the simultaneous occurrence of a certain psychic state with one or more external events which appear as meaningful parallels to the momentary subjective state. (p. 36)

To get to this place of understanding the meaning that underlies reality, we need to somehow overcome the natural bias we have when it comes to belief in an acausal principle. For a lot of us, including myself for most of my adult life, I would not have taken the scarab incident to *prove* anything mystical or meaningful. I, like many others, would maybe find it initially fascinating and interesting, but then would return to normal life; a kind of mental sweeping it under the rug. Somehow that's easier than having to accept a new worldview. But, for Jung, sweeping it under the rug and returning to a comfortable normalcy entails missing the entire point of existence.

So, how can we better have access to the level of meaning that true synchronicities are capable of providing us? How can we get past the skeptical blocks that are our natural tendency? For me, as a neuroscientist steeped in the scientific method and rational materialism, it has been a difficult road. It took a deep personal tragedy for me to begin to experience and pay attention to the immense significance of synchronicities, and to come to have faith in their legitimacy. If I hadn't had the personal experiences I had, I would most likely still be lacking in that faith.

Soon after my best friend and colleague of twenty years died, I began having very unusual experiences that made me suspect that he was somehow still around me. As a skeptic with lingering distrust of people who claim to be able to talk to the dead, this was—to put it mildly—out of my comfort zone. And yet, I was extremely curious. Could he really still be here in some way? I began to perform my own experiments to prove to myself he was. This was accomplished in a few ways. One was to ask him, upon waking up in the morning, for images in my mind that I would encounter later in the day. The first time I did this I saw an image of an old Wright Brothers plane. When I went running later that morning, I looked up at the rooftop of a house I had run past many times and there, attached to the roof, was a small replica of exactly that type of plane. While this was impressive, I was still able to question whether the information came from him. Maybe I'd subconsciously seen that plane but forgot, and I put that image in my mind myself. I had other experiences like this one but I was still not convinced. To be more sure, I tried a different method: asking for specific signs from him to let me know he was still here.

I was overwhelmed by the response. There were so many shocking moments where I would ask for something particular to be placed in front of me and I would receive it. Still I was hesitant to shake my materialist way of understanding the world. But one chain of signs I received finally resulted in the relaxation of my logical mind and allowed myself to jettison the last vestiges of doubt: I was driving north from San Diego to the Bay Area to visit my father in hospice. While on the highway, I asked my dead colleague to send me a sign that he was around me. I told him it didn't matter what the sign was, just something I would understand.

Just after I asked, I heard some clicking sounds, then the song "Hello it's Me" came on the radio. That jarred me. Was he saying hello? I wasn't sure. Needing more evidence, I asked for further signs. Just then a car passed me with a license plate that said Camelot on it, where instead of the 'e' there was a heart. This was actually very significant to me for two reasons. First, I knew that Camelot was associated with King Arthur. This reminded me of the nickname (Wort) that my father—who loved the Arthurian Legends—gave me, which was the nickname Merlin gave King Arthur when he was a child. And second, it also reminded me of the last time I had visited my colleague, and he took me to his favorite baking shop, "King Arthur's flour." This was all deeply meaningful to me, but I wanted yet more proof. So to clinch it, I asked him out loud, "if that was you I want the next song on the radio to have the word 'knight' in it." Amusingly, in my mind I was thinking of the song "Knights in white Satin," (amusing because later I came to know it isn't "Knights" but "Nights" in white satin). So there I was, waiting for "Nights in white satin" to miraculously play on the radio, but it didn't happen. Instead, the *very next* song that came on was a song by Neil Young that I hadn't heard for many years, called "After the Gold Rush." As soon as it started playing, I had a chill and a knowing: the first line of the song is "Well, I dreamed I saw the *knights* in armor coming sayin' something about a queen."

I can't describe the feeling I had at that moment; that absolute thrill of discovery and understanding; the sense of expansion and excitement. This was not something I could ignore or deny or sweep under the rug. More came, but it felt like icing on the cake. I passed two trucks on the way that were "Knight Transportation" trucks. Then, when I arrived at my family home, two books were placed facing out on the bookcase of the room I was sleeping in: both had "Knight" in the title. Next to those two books, also facing out, was Richard Feynman's book "The Meaning of it All." At that point I was floating, somewhere between our world and theirs, feeling the magic and wonder I didn't allow myself to previously feel. No amount of rationalization could penetrate this new knowledge. The probability of these chains of signs—and the specificity of the signs after having asked for them—being due to mere coincidence was astronomically low. As a scientist utilizing statistical methods to show the likelihood of hypotheses, I knew the likelihood that what occurred was due to normal causation was at least one in a trillion. Most likely much, much more improbable than that. And as a good scientist should, I accepted that my working hypothesis that my colleague was sending me a message, indicating he was still somehow 'here,' was validated.

I consider these types of experiences to be the epitome of synchronicity, as Jung would have us understand it. Not only the improbability of this chain of events, but the deep meaning behind them, both with respect to my deceased friend and my soon to be deceased father, was profound. But I also understand that anecdotal evidence is very hard to accept. I wouldn't have accepted it if it hadn't happened to me. The implications are too great, if wrong. It opens up the possibility of there being a universe that is not governed by rational laws as we know them. It opens up the possibility of parapsychological phenomena being true.I deeply understand the implications of allowing oneself to open to a vast, seemingly nonsensical reality. It's as if the floor has been taken out from under us, everything we used to 'know' thrown aside. But I came to understand that this can also be a great opportunity; an opportunity to willingly suspend our disbelief for a moment and consider the other possibilities, and whether those possibilities are worth the risk involved. There appears to be a choice here that we should be considering prior to dismissing Jung's principle out of hand.

Is there a way to help with making this choice? Blaise Pascal, a brilliant French mathematician and catholic writer living in the 17th century, grappled with how to make belief in God a rational choice. When I first started studying philosophy, I was taught about Pascal's famous wager. This intrigued me, as I was brought up in a religious household but rejected organized religion from a very early age, feeling even then that 'faith' wasn't good enough a reason to believe in something. Pascal provided a kind of calculus, or cost benefit analysis, in addition to faith to help those who require more. He provided a matrix of possibilities having to do with belief in God, and asked us to consider these four options: (1) God either exists or (2) doesn't. And we either (3) believe in God or (4) don't believe. The ramifications of the options, according to Pascal, are shown in the table below.

		We believe in god	
		Yes	No
God exists	No	Wasted time and effort	Time spent usefully
	Yes	Eternal salvation	Eternal damnation

Here it seems clear that the 'rational' choice is to believe in God, whether or not God exists. Weighing the pros and cons, the decision became obvious to Pascal. Some wasted time and effort is not worth eternal damnation. While Pascal's wager had a definite fearsome catholic spin, and may not seem applicable to many of us, we can still use it as a way to delineate the possible scenarios under our purpose of belief in Jungian acausal synchronicities. Assume our decision matrix consists of the following: (1) synchronicities exist, (2) synchronicities don't exist, (3) we believe in them, and (4) we don't believe in them. How might we fill in our matrix?

First, there is the situation where we believe in synchronicities but they don't exist. What is the cost of believing that 'coincidences' are meaningful when in fact they are not? This may lead to a life of succumbing to some form of magical thinking when there is no magic. We may suffer ridicule from our peers, or worse, from ourselves. This is the situation that, as an academic, I used to be very fearful of. Belief in something 'nonsensical,' or 'irrational,' may indicate that I'm less intelligent than my peers, or at best mentally unstable, which for many academics is to be preferred over lack of intelligence. Many great thinkers, even Jung himself, struggled with the fear of irrationality.

Now, what happens if synchronicities don't exist and we don't believe in them? We may have a sense of being vindicated, knowing we haven't been gullible and have correctly adhered to a rational state of mind.

But let's assume synchronicities exist and we *do* believe in them. This could lead to a deep understanding of reality where we have tapped into something beautiful and meaningful, providing our lives with a depth of knowledge and understanding we couldn't have otherwise.

Finally, if synchronicities exist and we don't believe in them, this may result in our having lost access to the true meaning of our lives.

It's one thing to be told that the rational choice is to believe in something; it's another thing to actually believe it even if you want to. When asked how one is to make oneself believe if they don't already, Pascal's advice was to fake it at first. Spend time with religious people, pray to God, act as if it's true and ultimately it will become second nature. Perhaps we can take his advice and 'fake' it at first, and see what happens. One way to get started is to begin to properly research and do experiments of your own, like I did. Understand that dismissal of a hypothesis out of hand is anti-scientific. If you really want to know the truth, some dedication to rigorous research on the topic is required.

It's ironic how many materialist scientists have not actually researched the ideas that they are lambasting. Hypotheses concerning acausal phenomena such as synchronicity, ESP, telepathy, mediumship, psychic abilities of any sort, for this type of person, are not worthy of proper analysis. They are not worth analysis at all. To them, they are wrong, *a priori*. As a scientist, I find this unacceptable if we are truly searching for truth. So jump in and at least make an attempt at discovery, even if it's difficult at first. Addressing the possibility of synchronicities in a logical, rational way might just lead you to discovering they exist and, thus, provide you with the beginnings of a much more meaningful existence. Open your eyes and mind to a new understanding and see what Jung understood, that "synchronicity is an ever-present reality for those who have eyes to see."

Reprinted with permission from Essentia Foundation (essentiafoundation.org) (7-2-23)

Chapter Twenty-Three

When Inner and Outer Align

LISA MILLER, PHD

Lisa Miller, Ph.D., is Professor of Psychology and Education at Columbia University, Teachers College and is Founder of the Spirituality Mind Body Institute, the first Ivy League graduate program in spirituality and psychology. Dr. Miller is a foremost scientist on spirituality across the lifespan, with her work published in top research journals including *JAMA-Psychiatry, American Journal of Psychiatry,* and the *Journal of the American Academy of Child & Adolescent Psychiatry.* Her innovative research has focused on quantifiable effects of spirituality in health, resilience and thriving, and an overall sacred and joyful life. Her clinical and consultation work focuses on spiritual awareness and spiritual growth, for individuals, families, groups and organizations, to augment "Awakened Awareness." She is the NY Times best-selling author of *The Spiritual Child*, and the author of the recent bestselling book, *The Awakened Brain.*

In the paragraphs below, I will share a few of the synchronicities in my life, though there have been many. And then I'll explain the significance of synchronicities for me, and what they say about the nature of reality.

Synchronicity 1.

Midway through my postdoctoral fellowship, my husband Phil and I decided to have a child. Parenthood had always been part of our plan, the life vision we shared as we fell in love, got married, and launched our careers. Our schooling done, careers in motion, finances steady, our fourth decade of life begun, the time was right. It wasn't so much a decision as a mutual embrace of the next logical, lovely thing. But months passed, and I wasn't getting pregnant. Each cycle, we rode the crest of hope and then the trough of disappointment. After a number of missed attempts, my doctor ran the standard tests. Nothing concerning showed up. There was no biological reason we couldn't conceive. Two more months passed, and still no baby.

"There are interventions to consider," my doctor eventually counseled. She handed me a stack of glossy, pastel-colored fertility brochures.

"Of course we can get you pregnant," the doctor said.

I'd lost count of the times a respected practitioner at a top-notch facility had said these exact words. I tried to ignore the feeling in my gut that he was wrong, that every time I went to an IVF clinic hoping to become a mother, I was knocking on the wrong door. It wasn't just my intuition telling me to stop. It was my body, too. I'd had so many injections that my stomach was swollen with bruises, and I was concerned that all the hormones would make me sick.

And yet, there was no physiological reason anyone could find that explained why I wasn't getting pregnant, and so I told myself not to lose hope ... I signed the forms and waivers, I took my drugs, and the doctor implanted an embryo in my womb.

I had to be on bed rest, and Phil stayed in the hotel room with me in solidarity. We were at a beautiful place in Rittenhouse Square, in a room that looked out on a peaceful park. We cozied up, ordered food, and looked for something good on TV. But when Phil tried to flip through the channels, the remote wouldn't work. We were stuck on a channel showing a depressing documentary shot in a garbage dump in Rio de Janeiro. A little boy, an orphan, stood on a heap of trash, interviewed through a translator.

"I don't care that I can't go to school," he said. "I don't care that I live in the garbage dump. But it hurts so much to not be loved that I sniff glue to make the pain go away."

I caught my breath. I looked at Phil.

He said it first: "There's a child out there for us."

I didn't know what seeing the orphan on TV meant, or if anything outward would change in our quest to become parents. I still desperately wanted to become pregnant. I still struggled with uncertainty and discouragement. But it also felt significant to have heard the orphan's story at that time. It gave me a flicker of hope, a recognition that the boy and I were part of the same reality, the same whole.

When the IVF treatment failed again, the familiar despair closed in. But a new feeling threaded through the dark—a sense of curiosity about how things might unfold.

One afternoon later that year, I came home from yet another IVF treatment, heavy with the familiar feeling that it wouldn't work. As I neared the front door, head foggy and out of sorts, I happened to glance down and see something on the doorstep: black, wet, the size of a finger. I stooped to examine it. A beak and tiny webbed feet dangled askew. It was a duck embryo. I used a piece of mail to scoot it gently onto the earth beside the door.

I got into bed with my clothes on and took a long, depressing nap. I dreaded Phil's return home and the next OB appointment when yet again we'd likely learn that the embryo had not implanted—that it had failed to find a home. I finally woke to a persistent tapping, loud and clear over the racing river. When I peered out the front window, I saw a full-grown duck, a female, thrusting her beak at the door. Tap, tap, tap. I opened the door and found that the mama duck had brought me a gift: a plump, juicy worm. She dopped it on the threshold and waddled back toward the river.

In that moment, my inner life and outer life lined up in a way that felt significant, too improbable to have happened by chance. I felt guided by something, a larger order or life force. In that moment, I saw the mama duck as evidence of the deep connection possible between living beings, a feeling of oneness. Even hope.

After the first duck-knock on the door, more synchronistic moments arose. I was on a bus in New York and, out of the blue, a stranger next to me said, "You seem like the type of mother who would go all around the world, adopting all kinds of kids."

A few days later my mom called. She's a frequent volunteer in my parents' suburban Boston community, and she had been working on a project with someone who hadn't been pulling her weight.

"Normally, I wouldn't say anything," she told me. "I'd just keep doing my part. But I was doing ninety percent, and it was too much, and I was about to confront her. But then she tells me she's preoccupied because she's just adopted a baby from Russia. She invited me over to meet him. Robert Abraham. Her beautiful little boy, as healthy as can be. Such a darling."

Synchronicity was building on synchronicity. It was as though perceiving synchronicity once, with the orphan on TV, and then again with the mama duck, had opened the way to perceiving more and more synchronicities. I had the sensation that, just as in an archetypal quest, I was becoming open, ever so gradually, to the guidance of helpers and healers on my journey. I was becoming aware of what hikers on the Pacific Crest and Appalachian trails call "trail angels"—the people who offer food and help and respite to weary hikers. I still had a long journey ahead, full of unknowns. But I didn't feel anymore that I was alone. I could sense that I was being held with support and caring by others along the path.

When I saw the duck embryo and received the worm, I didn't get pregnant the next month. Nor did I discard my desire to be a mom and decide that my life would be better without kids. I woke from a depressing nap and was granted a sudden glimpse that I'm part of a deep interconnectedness. That I'm being buoyed up, that I'm in the right place, I'm on my path, I'm part of the field of life. I was suffering, wondering if I could go on without a child. And then I heard life essentially telling me, *There's so much love and relationship and interconnectedness all around us. You are part of the oneness of life.* Depression wasn't the thing that had obscured my view of the bigger picture. It was the knock on the door.

Later in 2000, a few months after finding the duck embryo, I sat with Phil in a small room of a Pittsburgh adoption agency, every wall plastered with photos of smiling families from all over the United States who'd adopted babies and children from Russia ... You need to tell me honestly, " the woman from the agency said, "What do you want in a child?" Sophie Gurel leaned right up to me with a frank, piercing gaze ... The one element evident in every single photograph was love. It struck me: the definition of parenting is love; to be a parent means to love in a new capacity. I'd been so caught up in the goal of finally having a child—or, really, just getting pregnant-that I had lost sight of the larger view; to be a parent is to live the greatest love story.

"I don't care what race this child is, " I said. "I don't care about gender. But please, I want a child who *can love."*

She looked at Phil. "What about you?"

"Everything that Lisa said is true for me. But ..." He grinned sheepishly. "But, I'd kind of like a girl."

A few nights later, I had a profound experience. I suddenly woke out of a deep sleep. I sensed a sort of charged brightness in the room. And I felt something I can only describe as a presence there with us. My heart started racing. I sat straight up in bed. The presence spoke to me. I didn't hear it exactly. I sensed it, felt its resonance. *If you were pregnant would you adopt?* the powerful presence asked.

"No," I said into the night. My answer had come right out of my mouth before I could consider the question. But I knew it was the truth ... As soon as I spoke, the presence was gone.

My cousin Jane later called me at home and said in her frank, down to earth way, "If you're still trying for a child—and if you really want to understand what's going on—you need to come out here. Tomorrow." A clinical social worker and therapist, she sometimes worked with a nearby community of Lakota healers who would be gathering for a daylong ceremony. thinking they might be able to guide me in ways that Western medical science hadn't, she'd requested and received permission for me to attend with her ... Was I open to a different kind of intervention, something far outside my training as a scientist? I took the phone outside and stood barefoot on the deck, watching the river tumble over the rocks, evening light caught up in the tumult. "I'll come," I said.

When I stepped into the domed tent made from willow trees and draped with animal hides, I felt awkward and displaced. A fire burned inside, the heat was intense, the tent full of smoke. We sat in a circle on a bed of sage. Jane and I were the only non-Lakota women there. Everyone else wore jeans and blouses or shirts. I was dressed like I was going to brunch, in black pumps and a skirt with white polka dots. The heat was almost unbearable. Near the base of the tent was a small crack where a tiny wisp of cold fresh air came in, and I tried to inch myself closer to the gap.

The heat flared. Each woman was invited to speak in turn, to say why she had come, to share her suffering and need for healing. One woman spoke of her son. He was in his forties, grappling with addiction. He had stopped coming home. So many women shared about their sons—sons who were sick or hurt or disappearing. The sharing moved in a circle. I was the very last. My cousin Jane was on my left. When the circle of sharing reached Jane, I could feel her understand that I had no words. She spoke for me.

"I am Big Jane, and this is my cousin Lisa Jane," she said. "She has come searching for her child. Can we help her find her child?"

The women in the dark around the fire looked at me with deep attention. They nodded, they *mmm'd.* Their presence felt solid around me, sustaining. Here, the need for healing was not synonymous with brokenness. It was part of life. And these women understood the experience of looking for your child.

The medicine woman recited a prayer in Lakota. The others joined in. I felt the unity of the prayer, a presence coming together at our center. It gathered at the fire and then shot up, *wooosht,* opening a space. There was almost a light to it, a wind, directed up by the collective. The energy seemed fueled by an amalgam of our prayers, a oneness.

We left the tent, heading into the cool late night, the fresh clean air, under an open sky bright with stars.

As I left the next morning, I dialed into our voicemail machine. There was one message, left late the previous night, after Jane and I had left the inipi. My heart raced. Was something wrong with my parents? Or Phil's? Why hadn't he called me? Throat tight, I pressed "1" to hear the message, barely breathing as I listened to an unfamiliar woman's voice. It was Sophie, the rabbi's daughter, calling from Russia.

"We have found your child," she said. A baby boy, six months old, soon of legal adoption age in Russia, had been located at an orphanage in St. Petersburg.

Some weeks later, we receive a video of a joyous little baby body, happily gesticulating, raising his arms, smiling at a nurse. *"Da, da,"* he cooed in Russian. Yes, yes. He was pure love. He radiated it, a high-voltage laser love. A soaring euphoria consumed me, a love so immediate and powerful it moved through me like a tide, miraculous. I was carried and held.

Phil and I went to bed that night as parents. In the deep of the night, the presence came again—the full sense in the air, the numinous light. *If you were pregnant now,* it asked, *would you still adopt?*

"Yes!" I said. Tears brimmed my eyes. Of course I would still adopt. I had already met our son. I was madly in love with him. "Yes!" I cried again.

Phil stirred and woke. I embraced him.

"Let's name him Isaiah," I said. It's a Hebrew name; it means *Yahweh (God) is salvation.* "Isaiah Lakota."

"Yes," he said, laughing, kissing me.

That night we conceived.

Synchronicity 2.

When Leah and Isaiah were young and I was newly pregnant with our third child, Lila, Phil and I hired a babysitter one afternoon and headed out for a lunch at the local diner. It was February, cold and austere, the trees bare, dirty snow plowed into hills on the sides of the road. About a mile away from home, at a tricky intersection, I saw three young deer in their first winter, clustered tightly together, starkly staring, no mother to be seen. I felt an immediate connection with and concern for the deer, three lost children. Worried that they didn't know what to eat, or that there was simply not enough food on the branches, I sent them a message from my heart, mentally telling them, I will feed you. My achieving brain quickly retaliated, chastising, *How will they ever find you?* But the next day, there they were, all three arriving in my driveway at the exact moment I walked outdoors into the cold. They stared right at me. We had connected in the field of the awakened heart. I strongly suspect that the connection could have been picked up as alpha in an EEG, either on them or on me.

What kind of universe makes synchronicity possible?

As a researcher I understand that there was nothing in clinical science that accounted for what I was experiencing personally and clinically. There was no model that adequately described or explained the lift and ease I felt when my inner and outer life synced up—a buoyancy that came not from getting what I wanted, but from suddenly seeing my place in the world anew. Psychology held what seemed too limited a view on healing: that we scan the world and knit together meaning; that we feel better when we are able to choose to make a brighter meaning from the stuff in our lives. What I was witnessing in my fertility journey and in treating patients was less that we heal when we impose a more positive meaning on the world, and more that we shift toward health when somehow, and usually through struggle, a bigger meaning is revealed to us. Synchronicity seemed to be one way, even in darkness and suffering, for a new sense of the world to show up or shine through.

And synchronicity isn't just a matter of perception or awareness. It's also a physical phenomenon—something that actually happens in the world. At a moment of personal loss, I really had found a duck embryo on my doorstep, and I'd found it on that particular day, not weeks or months or years in the past or future. A mama duck really had brought a worm to the door. A stranger on a bus who couldn't possibly know anything of my struggle had mentioned international adoption. Many of us have these experiences. We might suddenly think of someone we haven't seen or spoken to or even thought of in ages, and the next day we bump into him on a crowded street.

But what causes synchronicity to occur in the first place? What makes our inner and outer realities align in what seems like random, improbable coincidence?

Could quantum physics offer an explanation or guidepost for how and why synchronicity occurs? If particles are entangled-interconnected and correlated across vast distances-might humans, made up of many particles, be intimately and inextricably entangled, too? And might synchronistic experiences be those moments when a wave of possibility is discernible as a particle?

And could quantum physics provide a new paradigm for psychology? A way to perceive and navigate a reality in which we are both particles and waves?

One of the things I've noticed about synchronicity is that it's about noticing and aligning ourselves with whatever life is showing us. We're given a sudden glimpse that we are a part of a deep interconnectedness. That we're being buoyed up, that we're in the right place, we're on our path, we're part of the field of life. We hear life telling us, *There's so much love and relationship and interconnectedness all around us. We are part of the oneness of life.* When we become spiritually aware—through synchronicity for example—it's a sign that despite the uncertainty in our lives, we are aligned with the force of life.

To identify, appreciate and then act on the guidance in and through the field of life, sometimes revealed through synchronicity, all of us have the innate capacity to awaken. We are built to be made whole when we perceive our alignment with the broader universe, which I share in my book, The Awakened Brain.

My takeaway? The questions far too often asked are: "What do I want, and how will I get it?" Instead, the real questions for an Awakened Life are: "What is life showing me now?" "What is the intelligent, loving Force of life, who I call G-d, asking of me now?"

Epilogue

How do we each perceive the inherent nature of life? Is its essence best portrayed by a deep sense of interconnected intelligence and order, embodying both purpose and meaning, or alternatively as a chaotic series of random chance occurrences over which we have little influence or control? In their essays in this volume scholars have confronted this question, often triggered through the exploration of the mysterious nature of synchronicity in their lives. These contributions examine a wide range of synchronicities and explore their implications concerning the nature of reality. In the following pages we will explore a few of the important themes that emerged within the essays, along with some additional thoughts of the authors regarding the nature of synchronicity.

Enchantment, disenchantment, reenchantment

In his book Breaking the *Spell of Disenchantment,* which describes his own exploration of synchronicity, one of our contributors, Roderick Main, begins by telling his audience about dropping out of formal education after his first year at the University of Oxford. He decided instead to live for a time in Glastonbury, in order to explore magic and write poetry. He said that within that first year of university life, he had felt that something vital was missing—a connection to the mystery, meaningfulness and sacredness of reality. He describes a feeling that many of us have encountered when entering the university environment. He was seeking enchantment, but at Oxford he felt he had entered a disenchanted milieu. Though his experiences during this first year at Oxford could be interpreted as a kind of ontological shock—a challenge to his worldview, After some time he did return to Oxford, and eventually pursued a PhD in Religious Studies at Lancaster University, going on to specialize in the exploration of synchronicity in his academic career.

When I, Marjorie, read these words of Roderick Main's I felt a great resonance, thinking about my own life journey, and how as a young university student in America, I had bowed to the disenchanting power of the materialist neuroscience teachings in my courses in Biology and Psychology. I relinquished my sense of enchantment and my search for the sacredness in life, taking on board the worldview of my professors as the only acceptable view for a young scientist. Unlike Roderick Main's journey, it was about 12 years before another very powerful and expansive experience opened me up to the possibility that the world was indeed enchanted. This event catalyzed the beginning of an exploration of this new worldview, and the synchronicities which become apparent as part of this new awareness. The journey from disenchantment to re-enchantment has continued to this day.

Through reading the essays, we have come to believe that many of our contributors, in their own way, have been following the path trailblazed by Jung in the 1920s when he began exploring the notion of synchronicity. He first used the term in a lecture in 1930 in the context of a discussion of Chinese religious concepts, for instance in consulting the *I Ching*. He developed this idea more fully in collaboration with the physicist Wolfgang Pauli, publishing his seminal essay in the 1951. Like Jung, the volume contributors were attempting to explore and more fully unfold the possibility of a worldview in which the universe was seen as sacred, and full of meaning and purpose.

We believe it would be helpful to give the reader a brief background on Jung's own journey from his sense of the disenchantment of the age of modernity, through his own efforts to reenchant the world by means of his own contemplations, writings and formulation of the concept of synchronicity.

About the time when Jung was beginning to explore synchronicity, the German social thinker, Max Weber spoke to a German student society, and specifically to students considering a scientific or academic career. He shared with them that in the past the quest for knowledge concerned a search for the true reality of being and said that this was no longer credible in the academy, as the principle of scientific and material progress now informed academic careers. He characterized this shift as the disenchantment of the modern world. In fact, the German word he used is *Entzauberung,* which means "de-magification" or "loss of magic.". In addition, it has been noted that, ironically, when the scientists of the age of "enlightenment" embraced the reductionist and materialist worldview, they may have thought they were freeing us from the naive enchantments of the pre-modern world, but in so doing they created an enchantment of their own, which has created a strangle-hold on our Western culture (Main, 2022).

Though Jung never used the term disenchantment in his own writings, he was clearly aware of this phenomenon described by Weber. In a lecture given at Yale University he addresses what he calls the historical process of despiritualization, and later describes the modern world as being desacralized, noting the extreme intellectual and rational nature of the modern age. He states, "We have stripped all things of their mystery and numinosity." (Main, 2022, p. 22). Jung's search for re-enchantment, through his writings on individuation (the developmental process of synthesizing the conscious and unconscious to become a unique self), and his development of the idea of synchronicity, have also been paralleled in the writings of others (e.g., Moore, 1966; Tarnas, 2006; Asprem, 2014; Main, 2022; Schwartz, 2017; Beitman, 2022).

Periods of crisis as a catalyst for synchronistic events

Jung noted that synchronicities often play a particularly powerful role during periods of crisis and transformation, appearing after a period of personal darkness. They may disrupt an individual's ontological security, defined as a sense of order and continuity in how an individual sees the world (Giddens, 1991). In this way, synchronicities deliver an "ontological shock", an occurrence of something that is beyond one's current understanding of reality, and forces one to question underlying beliefs related to the nature of reality (see Main, this volume).

A powerful example of this is the account of one of our authors, who, along with her daughter, was about to begin a new life together with her fiancé, Jerry, when he suddenly had a heart attack and died. She was devastated and concluded that if something as awful as this can happen then surely that is proof that there is no benevolent power giving meaning in life. However, something kept nagging at her, and so she spent considerable time contemplating this dilemma. She began to consider that perhaps there was some meaning to this awful event, but it was beyond her ability to understand, at least at that moment. She began to think "I need to remain open to the possibility that there is meaning in this event" and she created the hypothesis that his death was meaningful, that there was a purpose behind it, and that one day she would understand it. She would act as if this was true.

She says, "A number of years later, my daughter was driving her car over a bridge out in the country. Because it was a single lane bridge it was important to slow down and make sure that no cars were coming in the opposite direction. She slowed down and then tried to accelerate over the bridge, but her foot wouldn't move on the accelerator. At that moment a car came speeding over the bridge, not slowing down or looking for other cars and zoomed past her. Because of her inability to move her foot and accelerate her car over the bridge, her daughter had avoided a head-on collision ... as she drove on, she realised the significance of what had just happened. Her life had been saved, because her foot suddenly froze, and she couldn't press the accelerator. A strong sense of Jerry pervaded her consciousness at the same time. What possible explanation could there be for the concurrence of these two aspects of her experience, when she was so occupied with what was happening in the present moment, and no time to think and reflect? (Walton)

The author said, "It feels to me to be a strong affirmation that there is indeed a wider reality, which we are not able to see directly from a limited human perspective, but is indeed present with us, and is supporting us on this life journey, if we are open to that possibility."

In a second example, one of our student contributors describes the events leading up to her own exploration of synchronicity in her life, saying:

> I was nearing the end of my undergraduate degree in psychology and ...I was feeling deeply unsatisfied with the lack of depth in my learning experience so far. I hadn't felt transformed or enlightened as I had expected when I began college three years prior. Since my curiosity wasn't being satiated within school, I began to look outside the curriculum. I came across some books on quantum physics, and one book in particular, *The Self-Aware Universe,* by Amit Goswami (Goswami, 1995). I immediately felt something stir within as I began to understand how spooky and magical the implications of quantum physics are—as well as what this could mean for consciousness—that it is primary rather than an epiphenomenon (Kothe).

This period of crisis and the ontological shock she experienced as she explored the book, *The Self-Aware Universe,* helped her begin to recognize synchronicities in her own life. And shortly afterwards in a bookstore she had a powerful synchronicity leading her to discover Jung's book of that title.

One of our contributors explains these synchronicities by noting that research suggests that we live in an entangled universe, as also suggested by Dean Radin (Radin, 2006). In his essay, Beitman states: "Synchronicity helps us to see the far reaches of our interconnectedness." This felt sense of interconnectedness is more than cognitive knowledge but comes from an emotional surge of connectedness. He reminds us that Jung called synchronicities "numinous" experiences, a momentary experience of oneness. Synchronicities, he suggests, may occur because what you are seeking is also seeking you.

This is echoed by Tarnas, who suggests that synchronicities have much in common with mystical experiences.

The probability of experiencing a synchronicity

We might ask if certain activities or events in life increase the probability of experiencing or being aware of synchronicities. Bruce Greyson performed a study to determine if the perceived incidence of synchronicities (called meaningful coincidences in the publication) was increased in individuals after a near-death experience. He found that coincidence scores on the WCS-2 scale (Weird Coincidence Scale-2) (Coleman et al., 2009) before the NDE were 2.43+.69 before the NDE, slightly below the population norm, and had increased significantly to 3.43+.84 after the NDE. Interestingly, participants with higher scores on the NDE scale (Greyson, 1983) were associated with higher scores on the WCS-2 scale, after the NDE, indicating that the deeper the NDE the more the person perceived coincidences in their subsequent life experiences. In addition, closeness to death during the NDE was positively associated with scores on the WCS-2 scale.

Atmanspacher and Fach (2019) also explored what they called coincidence phenomena in different population samples, including the general population, meditators and persons who had experienced an NDE. Frequencies were lowest for the general population and highest for meditators and subjects reporting an NDE. Thus, from the results of both of these studies, we conclude that persons experienced in meditation or having experienced an NDE, are more likely to perceive synchronicity in their lives. This may be the result of their intense experience of oneness and interconnectedness.

Transformation in worldview associated with synchronicity experiences

One aspect of the journey toward re-enchantment is the desire to formulate a new worldview that replaces the disenchantment of materialism. Jung, along with Wolfgang Pauli, worked to formulate a new model of reality, which he called dual-aspect monism (consisting of an undivided whole, which is neither mental nor physical, which produces mental (conscious) and physical (objects) domains).He saw this as a revival of the Hermetic Renaissance idea of the *Unus Mundus*. In fact, many well-known scientists, including physicists, neuroscientists, and psychologists, like Jung, have been working to break this stranglehold of disenchantment brought about by the reductionist worldview throughout the last few centuries. These include, among others, William James, Erwin Schrödinger, Wolfgang Pauli, and even more recently, the well-known neuroscientist, Christof Koch.

Koch's journey is especially interesting. As a young scientist, Koch and his collaborator Francis Crick stated that they would soon solve the hard problem of consciousness, that is, how neurons in the brain can produce consciousness. After many years of research, Koch has admitted that this was not possible. At the 2024 Science of Consciousness meetings at the University of Arizona, he stated in a plenary session that finding the neural correlates of consciousness was insufficient to satisfactorily explain consciousness, and this caused him to make a decisive shift away from classical physicalism. He concluded his talk by saying that now he leans toward some forms of idealism, as closer to the mark. Such a view has been systematically developed as analytic idealism by Bernardo Kastrup (Kastrup, 2019b). These are remarkable statements from someone who was firm physicalist for much of his academic career.

We note above that Jung's reflections on synchronicity led him to develop a worldview which he labeled dual-aspect monism. Given that Jung explored a new way of viewing the world related to his exploration of synchronicity, we asked the authors of the essays in this book to describe characteristics of their current worldview in light of their experiences (we applied a similar experiential framing in our 2022 volume *Spiritual Awakenings* highlighting the role of personal experience in the transformation of worldviews). It is interesting that all of the authors supported theories of reality in which consciousness was considered fundamental. Words they used to characterize this worldview were interconnected, entangled, an integrated whole, alive, immanent and transcendent, knowing itself, a numinous mystery, possessing boundless creativity, vast and limitless, and a loving awareness.

Comments included:

- We all have reached the conviction that there are many more things in heaven and earth than are dreamt of by the reductionist materialism characteristic of the mainstream modern mind-set. For those who have attuned their ears to hear and eyes to see, opened their minds and hearts to the subtle ongoing reality of synchronistic phenomena around them, events like this intimate a larger intelligent order at work in life—deeper purposes, boundless creativity, and above all, a sense of numinous mystery that underlies all our lives. (Tarnas)

- The idea that we have an eternal self, what all religions recognize, calling it the soul or the spirit, the élan vital, the animus, higher self, and through the centuries a host of other terms. I call it the eternal self, and the evidence from reincarnation, NDE, and other research suggests that this eternal self episodically manifests and incarnates a personality, and that what comes across incarnations is information. (S. Schwartz)
- Consciousness is the fundamental expression of the immaculate power of love, which is the initiator of its creation ... The One, multifaceted and multi-aspect reality that contains and is contained within the whole cosmos, the *Unus Mundus* (ἕν το πᾶν) of the Ouroboros of Jung and Pauli. It holds the interconnected diversity of its parts by being "within" those parts. (Basios)

Some authors identified specific theories, such as idealism, dual-aspect monism, or panentheism, as their ontological framework. However, their description of these theories largely overlapped. In fact, one of the essayists said of his own view, panentheism, "As an overarching framework, panentheism can also host, it seems to me, various other, more philosophical and scientific frameworks that have been deployed to make sense of synchronicity, such as emergent panpsychism (Cambray, 2009), Whiteheadian process philosophy (Haule, 2011), objective idealism (Kastrup, 2021), and dual-aspect monism (Atmanspacher & Rickles, 2022)—though this intuition still needs to be worked through in detail" (Main, this volume).

These transformations in worldview, from a mechanistic disenchanted reality in which science and spirituality are irreconcilable, toward a re-enchanted reality where science and spirituality are integrated, were typical of the contributors to this volume. However, if we left the exploration of re-enchantment at this simple level of analysis, we would not acknowledge the complexity of human individuation and spiritual growth that Jung discusses in his own writings. Human development involves a process of enchantment, disenchantment and re-enchantment that occurs repeatedly in the life of everyone wanting to know the truth of who they and the world really are.

One of our contributors noted that we cannot completely leave the technical world and its accompanying focus on the physical behind, but must instead find a way to enfold aspects of the materialist framework into a post-materialist world view, as one deepens one's understanding of the nature of reality. And when pursued throughout one's life, this process of enfoldment of stages of enchantment, disenchantment and re-enchantment may culminate in a state of spiritual integration, including an acceptance of the beauty and mystery of life as it is. This might be called *gnosis*, a state of union with the vast and limitless consciousness which is experienced as the ground of being (Main, 2022). As Meister Eckhart put it, "the eye with which I see God is the same eye through which God sees me; my eye and God's eye are one eye, one seeing, one knowing, one love."

One fact stands out amongst the diverse, interconnected, overlapping perspectives presented in this volume – that synchronicities are real and pervasive, and they provide a window into a vast and limitless cosmic consciousness whose qualities show evidence of extraordinary intelligence, creativity, and caring. As Tarnas shares in his Foreword to this volume, Jeffrey Kripal, the distinguished scholar and professor of religious studies once confided to him, "I have stopped believing in most things, but I do believe in synchronicity."

References

Atmanspacher, H. & Fach, W. (2019). Exceptional experiences of stable and unstable mental states understood from a dual-aspect point of view. *Philosophies 4,* 7. doi:10.3390/philosophies4010007.

Atmanspacher, H. & Rickles, D. (2022). *Dual-Aspect Monism and the Deep Structure of Meaning.* Routledge.

Beitman B. *Meaningful Coincidences: How and why Synchronicity and Serendipity Happen.* Park Street Press, 2022.

Cambray, J. (2009). *Synchronicity: Nature and Psyche in an Interconnected Universe.* Texas A & M University Press.

Coleman SL, Beitman BD, Celebi E. (2009). Weird coincidences commonly occur. *Psychiatr Ann. 39,* 265–270.

Giddens, A. (1991.) *Modernity and Self-Identity: Self and Society in the Late Modern Age.* Cambridge: Polity.

Greyson B. (1983). The near-death experience scale: Construction, reliability, and validity. *Journal of Nervous & Mental Disease*. 171.6, 369–375.

Haule, J. (2011). *Jung in the 21st Century, Volume 2: Synchronicity and Science*. Routledge.

Kastrup, B. (2019). *Decoding Jung's Metaphysics: The Archetypal Semantics of an Experiential Universe*. Iff Books.

Kastrup, B. (2019b). *The Idea of the World*. Iff Books.

Main R. *Breaking the Spell of Disenchantment: Mystery, Meaning and Metaphysics in the Work of C.G. Jung*. Zurich Lecture Series in Analytic Psychology, Vol. 8, Chiron Publications, 2022.

Radin, D. (2006). *Entangled Minds: Extrasensory Experiences in a Quantum Reality*. Paraview Pocket Books.

Schwartz, G.E. *Super Synchronicity: Where Science and Spirit Meet*. Param Media, 2017.

Tarnas, R. *Cosmos and Psyche: Intimations of a New World View*. Viking, 2006.

Woollacott, M., Lorimer, D. (2022) *Spiritual Awakenings*. AAPS Press.

The Scientific and Medical Network (Founded 1973)

WWW.SCIENTIFICANDMEDICAL.NET

Where evidence-based reason meets deep inner knowing

Our vision

The SMN is working with full awareness and appreciation of scientific method, but exploring and expanding, in a spirit of open and critical enquiry, frontier issues at the interfaces between science, health, consciousness, wellbeing, love and spirituality, to explore how to rediscover a meaningful spirituality to help rebalance our lives.

The Network is part of a worldwide contemporary movement for education, personal development, and compatible "spiritual emergence", networking: scientists, doctors, psychologists, educators, engineers, philosophers, complementary practitioners and other professionals, for mutual and societal benefit.

Raison d'Être

For almost 50 years, the Scientific and Medical Network (SMN), has been a creative force for education, learning, and transformative change, supporting professionals to see that relying solely on scientific method (though excellent and indispensable in its province), cannot fully guide humanity. The results from these misunderstandings and misapplications, are now so startlingly being seen on our environment, health and well-being. This appreciation is not new.

SMN approach

Heart Brain Coherence

We believe that a better, more intelligent and more heart-led world is possible—a world in which the wisdom of spirituality and the thoughtfulness of philosophy balance the power of science and technology.

Deep Listening, Dialogue and Mutual Trust

It is our view that no single idea or solution has a monopoly on truth. Our events and publications, therefore, create spaces for sharing different ideas and experiences with an ethos of deep listening, dialogue and mutual trust.

Integral Inquiry

We recognise that our world is still plagued by deep problems and enduring conflicts. Many of these are caused by dogmatic, superstitious or ill-informed worldviews. However, we believe that individuals and societies can transform themselves.

The key is learning to integrate evidence-based science, born of rational philosophy, with the love and inner peace cultivated by spiritual practice. The Scientific and Medical Network is a community of professionals dedicated to this task.

Co-creating New Paradigms

We invite you to join us in co-creating new paradigms to bring truth, beauty, health and well-being into harmony, and so help heal our fractured culture.

Galileo Commission

WWW.GALILEOCOMMISSION.ORG

Expanding Science beyond a Materialistic Worldview

The world today is dominated by science and by its underlying assumptions, which are seldom explicitly articulated. The Galileo Commission's remit is to open public discourse and to find ways to expand science so that it can accommodate and explore important human experiences and questions that science, in its present form, is unable to integrate.

Following widespread consultation with 90 advisers representing 30 universities worldwide, we have published the Galileo Commission Report, written by Prof Dr Harald Walach and entitled ***Beyond a Materialist Worldview—Towards an Expanded Science.*** The report has been widely endorsed as a groundbreaking document and we encourage you to read it for yourself and spread the word among your professional network. Summaries of the argument are available in a number of languages.

A more recent initiative is *A Call for a Renaissance of the Spirit in the Humanities.* This document argues for a return to a spiritual image of the human being capable of direct unitive knowledge or *gnosis* so that we know ourselves more deeply and as essentially identical with the One Universal Mind.

We invite scientists and academics to join our community of over 350 professional affiliates and members of the general public to support us as friends.

Index

B

C

D

E

I

K

Q

R

S

V

W

Y

Made in United States
Troutdale, OR
07/15/2026

51022112R00156